AF836363

MANUFACTURING TECHNOLOGY RESEARCH

PRINTED ELECTRONICS

TECHNOLOGIES, APPLICATIONS AND CHALLENGES

MANUFACTURING TECHNOLOGY RESEARCH

Additional books in this series can be found on Nova's website under the Series tab.

Additional e-books in this series can be found on Nova's website under the eBooks tab.

MANUFACTURING TECHNOLOGY RESEARCH

PRINTED ELECTRONICS

TECHNOLOGIES, APPLICATIONS AND CHALLENGES

KYLIAN YVON
AND
NATHAN FABRICE
EDITORS

nova
science publishers
New York

Copyright © 2017 by Nova Science Publishers, Inc.

All rights reserved. No part of this book may be reproduced, stored in a retrieval system or transmitted in any form or by any means: electronic, electrostatic, magnetic, tape, mechanical photocopying, recording or otherwise without the written permission of the Publisher.

We have partnered with Copyright Clearance Center to make it easy for you to obtain permissions to reuse content from this publication. Simply navigate to this publication's page on Nova's website and locate the "Get Permission" button below the title description. This button is linked directly to the title's permission page on copyright.com. Alternatively, you can visit copyright.com and search by title, ISBN, or ISSN.

For further questions about using the service on copyright.com, please contact:
Copyright Clearance Center
Phone: +1-(978) 750-8400 Fax: +1-(978) 750-4470 E-mail: info@copyright.com.

NOTICE TO THE READER

The Publisher has taken reasonable care in the preparation of this book, but makes no expressed or implied warranty of any kind and assumes no responsibility for any errors or omissions. No liability is assumed for incidental or consequential damages in connection with or arising out of information contained in this book. The Publisher shall not be liable for any special, consequential, or exemplary damages resulting, in whole or in part, from the readers' use of, or reliance upon, this material. Any parts of this book based on government reports are so indicated and copyright is claimed for those parts to the extent applicable to compilations of such works.

Independent verification should be sought for any data, advice or recommendations contained in this book. In addition, no responsibility is assumed by the publisher for any injury and/or damage to persons or property arising from any methods, products, instructions, ideas or otherwise contained in this publication.

This publication is designed to provide accurate and authoritative information with regard to the subject matter covered herein. It is sold with the clear understanding that the Publisher is not engaged in rendering legal or any other professional services. If legal or any other expert assistance is required, the services of a competent person should be sought. FROM A DECLARATION OF PARTICIPANTS JOINTLY ADOPTED BY A COMMITTEE OF THE AMERICAN BAR ASSOCIATION AND A COMMITTEE OF PUBLISHERS.

Additional color graphics may be available in the e-book version of this book.

Library of Congress Cataloging-in-Publication Data

ISBN: 978-1-53612-797-3

Published by Nova Science Publishers, Inc. † New York

CONTENTS

PREFACE

In the first chapter, Urška Vrabič Brodnjak, PhD, Deja Muck, PhD, Helena Gabrijelčič Tomc, PhD, and Igor Karlovits, PhD provide a review of the many facets of printed smart labels in packaging, including the categorization, materials, techniques, environmental impacts, applications, and future trends. In the second chapter, Maja Klančnik, PhD, Helena Gabrijelčič Tomc, PhD, Diana Gregor Svetec, PhD, Marica Starešinič PhD, and Tatjana Rijavec, PhD give an overview of electrically conductive materials, fibers, inks, and printing solutions. In the third and final chapter, Umit Y. Ogras, Ujjwal Gupta, Jaehyun Park, and Ganapati Bhat compare FHE systems-on-polymer with traditional systems-on-chip.

Chapter 1 - The presented chapter is an extensive overview of the categorisation, materials used, printing technologies, environmental impacts, applications, future trends and potentials of printed smart labels in packaging. In the first part, the smart labels are defined and their role in the wider context of communication media is explained. Besides the aims of the packaging of the future, the networking objectives and the connection with the paradigm of Internet of Things are also presented. Further, types of smart labels are categorised and classified in electronic article surveillance, radio frequency identification (RFID), smart labels for brand enhancement, smart labels for diagnostics, electronic shelf labels and hybrid electronics smart labels. The focus of the contribution is on RFID for packaging applications and therefore RFID labels with printed sensors. In the second part, the materials used in smart labels are analysed, including inorganic and organic materials and substrates. In the contribution of printing technologies for smart labelling the comparison of additive and subtractive manufacturing patterning is shown and the comparative overview of printing technologies (offset, gravure, flexo, screen, inkjet, μCP, nano-imprint and dispense) is presented including the systematic comparison of parameters as printing form, ink viscosity, line width and thickness, printing speed, etc. Furthermore, the application values of printed smart labels are analysed. The implementations of printed electronic with the focus on RFID and sensing devices are discussed, including already launched and successful commercial solutions and some products at developmental research level. In the final part of the contribution, the environmental impacts of smart labels are explained. The presentation of negative impacts does not overlook the fact that there is a difference between active and passive RFID and that the components of smart labels should be treated differently, especially according to their material and complexity of components. The environmental positive impact of the printed smart labels is demonstrated especially when considering many opportunities for the improvements and reorganisation of management processes of recycling systems. In the conclusions, the implementations of information processing and augmented reality in printed smart labels for packaging are discussed. Here, the conceptual frameworks for smart packaging are presented, challenging the research field with some considerations about the need of constant redefinition and clarification due to persistent changes in the interconnectivity and relationship between physical and virtual media.

Chapter 2 - The presented chapter is an overview of technologies, materials, applications and potentials of electrically conductive materials, electrically conductive fibers, conductive printing inks and printing solutions with conductive inks. The wider aspects of the role of conductive materials and inks in wearable technology is presented with the contributions about the environmental impacts of electronic textile wastes and future potentials and concerns about the interconnectivity of physical and virtual world in wearables. In the first part the readers are introduced with the fundamentals of electrical conductivity and resistance of materials. Here, basic definitions are presented, including definitions and equations for electrical conductivity, specific electrical conductivity, electrical resistance, sheet resistance and bulk resistance. Further, the focus is on the electrically conductive materials for the production of conductive textiles, i.e., metals, carbon and conductive polymers. In the next chapters fibres and electrically conductive yarns are shortly explained, with the analysis of the materials used, mechanical and electrical properties and costs of their production. The section of printing technologies includes the explanation of conductive inks, followed by the systematic overview and presentation of different printing technologies (gravure, flexographic, screen and ink jet). The technologies and their applications are also explained with the recently published researches and developments. The chapter continues with the environmental impact of printing ink and proposals for more environmentally-friendly printing solutions. For the understanding of the environmental problem of electronic textile wastes that the society will encounter is further presented the analysis of the production, market demands and waste policies of e-textiles and wearables. The possible solutions for minor environmental impact of electronic textile wastes are proposed with the detailed explanation of eco-design principles and possible solutions with favorable consequences for sustainable developments of e-textiles. The contribution concludes with the future potentials of the use of wearable electronics in textiles. In this part the e-textiles are presented by means of their fundamental role, i.e., the human second skin that should not overlook the basic needs, desires, experiences and expectations of the humans. With some observations about the human sensory system, cognition and the interpretation of the data acquired, measured and processed by e-textiles, the authors predict a new paradigm of design and user-experience of wearables.

Chapter 3 - Flexible electronics have the potential to revolutionize the way humans interact, communicate, and control electronic devices. Cost and weight advantages combined with physical flexibility open a wide range of application areas, including wearable electronics, medical devices, and internet of things (IoT). However, performance and capabilities offered by flexible electronics technology are severely limited, and far lower than traditional rigid systems-on-chip (SoC) used today. Emerging flexible hybrid electronics (FHE) target this problem by integrating traditional rigid ICs and printed electronics on a flexible substrate, such as polymer. FHE can be used to coalesce rigid and flexible resources judiciously to drive the next big leap forward in the form factor design, similar to the shift from desktop and laptop computers to hand-held devices. This chapter compares the FHE systems-on-polymer (SoPs) with traditional SoCs and introduces the physical flexibility as a new dimension in the design space. It overviews the major requirements of flexibility-aware design methodologies and tools tailored for FHE. Finally, it discusses near- and longterm application areas and relevant challenges.

In: Printed Electronics
Editors: K. Yvon and N. Fabrice
ISBN: 978-1-53612-797-3
© 2017 Nova Science Publishers, Inc.

Chapter 1

PRINTED SMART LABELS IN PACKAGING

Urška Vrabič Brodnjak[1], PhD, Deja Muck[1], PhD, Helena Gabrijelčič Tomc[1], PhD, Igor Karlovits[2], PhD, Diana Gregor-Svetec[1], PhD and Marica Starešinič[1], PhD

[1]Chair of information and Graphic Arts Technology, Department of Textiles, Graphic Arts and Design, Faculty of Natural Sciences and Engineering, University of Ljubljana, Ljubljana, Slovenia

[2]Pulp and Paper Institute, Ljubljana, Slovenia

ABSTRACT

The presented chapter is an extensive overview of the categorisation, materials used, printing technologies, environmental impacts, applications, future trends and potentials of printed smart labels in packaging. In the first part, the smart labels are defined and their role in the wider context of communication media is explained. Besides the aims of the packaging of the future, the networking objectives and the connection with the paradigm of Internet of Things are also presented. Further, types of smart labels are categorised and classified in electronic article surveillance, radio frequency identification (RFID), smart labels for brand enhancement, smart labels for diagnostics, electronic shelf labels and hybrid electronics smart labels. The focus of the contribution is on RFID for packaging applications and therefore RFID labels with printed sensors. In the second part, the materials used in smart labels are analysed, including inorganic and organic materials and substrates. In the contribution of printing technologies for smart labelling the comparison of additive and subtractive manufacturing patterning is shown and the comparative overview of printing technologies (offset, gravure, flexo, screen, inkjet, μCP, nano-imprint and dispense) is presented including the systematic comparison of parameters as printing form, ink viscosity, line width and thickness, printing speed, etc. Furthermore, the application values of printed smart labels are analysed. The implementations of printed electronic with the focus on RFID and sensing devices are discussed, including already launched and successful commercial solutions and some products at *developmental* research level. In the final part of the contribution, the environmental impacts of smart labels are explained. The presentation of negative impacts does not overlook the fact that there is a difference between active and passive RFID and that the components of smart labels should be treated differently, especially according to their material and complexity of components. The environmental positive impact of the printed smart labels is demonstrated especially when considering many opportunities for the improvements and reorganisation of management processes of recycling systems. In the conclusions, the implementations of information processing and augmented reality in printed smart labels for packaging are discussed. Here, the conceptual frameworks for smart packaging are presented, challenging the research field with some considerations about the need of constant redefinition and clarification due to persistent changes in the interconnectivity and relationship between physical and virtual media.

Keywords: printed smart labels, packaging, RFID, printing technology, application, environmental impact, augmented reality

1. Smart Labels

A smart label is an identification tag or slip which enables more sophisticated data communication than a conventional bar code. It can be applied to various types of labels and tags, which generally share the characteristic of having flat responsive electronic elements (printed, silicon bases or hybrid) which can be connected to any network. In this way, packaging and various other items are transformed into smart objects, and can therefore be one of the future key enabling technologies for setup of the Internet of Things (IoT) (Briand et al., 20111, Jongwoo et al., 2015).

The smart label commonly lies under the print-coated label, like an inlay, and includes an antenna, chip, and bonded wires. The labels are typically made from fabric, plastic, or paper and are used as chip labels, printable labels, and electronic labels.

There are many types of smart labels, including (IDTechEx, 2004, Gregor-Svetec et al., 2015):

- **Electronic Article Surveillance (EAS)**, which are used for theft detection in shops (Figure 1) and libraries. These labels do not contain data and are fitted in packaging, so a prospective thief does not know they are there.

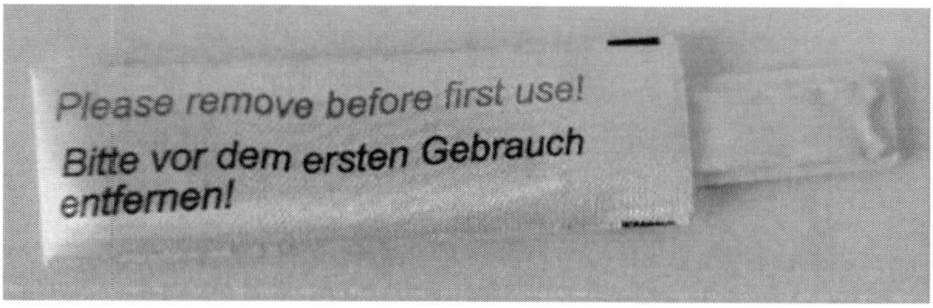

Figure 1. Electronic article surveillance tags for theft detection in clothes and fashion shops.

- **Radio Frequency IDentification (RFID)** smart labels are identification (ID) tags used in the supply and usage chain (Figure 2) by a range of stakeholders from parts suppliers, product manufacturers, retailers and repair services to consumers and recyclers. They are extremely widespread and can operate in different frequency ranges. They are also used in healthcare, pharmaceuticals, to prevent crime, for entertaining, refreshing brands, postal services, the military and archiving, including in libraries.

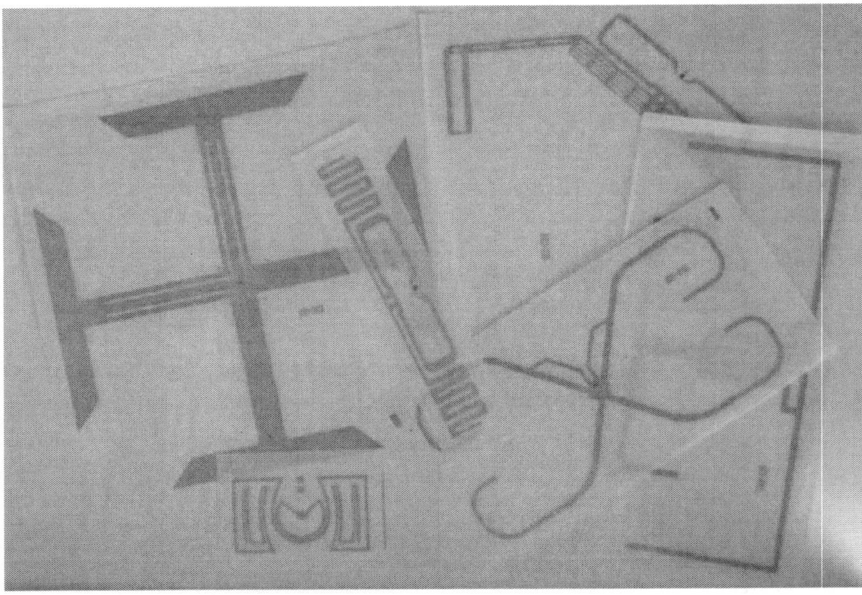

Figure 2. Ultra high frequency (UHF) smart RFID labels for supply chains.

- **Smart labels for brand enhancement** are labels on which a brand enhancement attribute (via a label or insert) is added as a useful feature directly to a product or its packaging. Examples are disposable battery testers.
- **Smart labels for diagnostics** are used on packages to detect overheating, shock, tilt, vibration, or completion of sterilisation. Such labels can be fabricated using non-electronic inks and laminates or functional electronic inks. While non-electronic versions are low-cost, electronic versions can signal what is happening in real time and most can store a digital record; however, the prices of the latter are higher.
- **Electronic shelf labels (ESL)** or dynamic display labels enable automatic information updates, such as a pricer's ELS (Figure 3). They are more cost-efficient than conventional paper labels, and improve customer experience by providing accurate, timely and precise information.

Figure 3. Electronic shelf smart labels.

Nowadays, smart labels are becoming more and more complex by creating intelligent elements on the surface of the packaging. With combined functionality, one label can perform two or more of the above categories of smart function, such as EAS + RFID (NFC), RFID (NFC) + sensing, and more. New technologies include polymer film electronics, microsensors, microbatteries, ultra-small microchips, various printed sensors and batteries, and flexible hybrid electronic technology.

Today's smart labels can be classified into the following categories:

- integrated components (battery, transceiver, memory, microprocessor);
- operation technology (Near Field Communication, Electronic Article Surveillance, RFID and sensing label);
- final application (equipment, IT and electronic assets, retail inventory, pallets tracking, perishable goods);
- end-users (healthcare, logistics, manufacturing, fast-moving consumer goods, automotive industry).

The smart label market is witnessing significant and healthy growth, and is expected reach a CAGR (*Compound Annual Growth Rate)* of approximately 17% during the forecast period (2015 to 2019) (IDTechEx, 2004). In terms of technology and value, RFID smart labels hold the largest market share.

In this chapter, the authors have focused mainly on printed RFID smart labels with internal and external sensors for packaging applications.

1.1. Printed RFID Smart Labels

Radio frequency identification (RFID) smart label is in its simplest form composed of an RFID tag with an antenna and an integrated circuit (IC), which takes the form of an inlay placed between the print-coated label and the adhesive substrate (Figure 4).

Figure 4. RFID smart label with tag inlay.

Depending on the communication method (active, passive or semi-passive) and operating frequency (low (LF), high (HF) or ultra-high (UHF)), the RFID tags for applications in smart labels can be different, and the reader and antenna need to be tuned to the tag frequency.

The most popular RFID tags are low-cost passive tags that draw energy from the electromagnetic field radiating from the reader. RFID tags can be produced using screen printing, gravure, flexography, and inkjet printing.

Of the different operating frequencies, UHF RFID tags and smart labels are the easiest to print, as their antennas can be designed as a flat one-layer conductive printed structure.

Over the past years, a substantial amount of research has been conducted into printing HF and UHF tags on different substrates, such as cardboards, foils and even recycled papers. The properties of recycled papers and cardboards required to produce printable antennas have been analysed (Kavčič et al., 2013, Kavčič et al., 2014, Gregor-Svetec et al., 2015). In one research paper (Kavčič et al., 2015), researchers successfully printed UHF RFID tags directly onto cardboard for the fabrication of pharmaceutical packaging. They found out that all samples could be successfully read when both horizontally and vertically oriented, although none of the tags could be read successfully when placed in close proximity to a reflective metal surface.

The effect of liquid in glass vials is less pronounced than the effect of reflecting metal, and tags can be read under these conditions, albeit with a shorter reading range.

In other research papers (Đokić et al., 2014, Đokić et al., 2015) researchers worked on an RFID tag inlay used for smart card applications. As with smart labels, the lamination process can have a significant impact on operability, especially reading range. They analysed the characteristics of screen and flexographic printing and the lamination process for RFID applications. The characteristics of the antenna were analysed by measuring several parameters such as resistance, backscattered power, the characteristic impedance, and the return loss. The effects of the lamination process on the final cards' readability and operability were analysed in detail. Possible printing processes that might be used in the realisation of RFID antennas were investigated in terms of their fabrication speed, costs for mass production, and repeatability of the characteristics. The researchers proved that industrial screen printing and flexographic printing techniques can be used with equal effectiveness for producing both RFID smart cards and smart labels; however, flexographic printing proved to be the faster and more cost-effective solution.

Nowadays, more and more smart labels have integrated various sensors, and we are talking about sensing devices or sensing systems. On printed RFID smart labels, antennas are mainly printed, but external sensors are Si-based.

There are some approaches to designing sensing systems based on passive, semi-passive or active RFID tags for packaging and logistics applications (Unander and Nilsson, 2009, Nilsson et al., 2012). Some of these designs combine microcontroller architectures with RFID chips to incorporate different environmental sensing devices, including: temperature, light, and acceleration (De Donno et al., 2014); temperature, humidity, and various gases (Oprea et al., 2009, Smits et al., 2012, Pursula et al., 2013); various ion or gas concentrations (Steinberg et al., 2014,) Kassal et al., 2013, Eom et al., 2012); and pressure (Fernández-Salmerón et al., 2015, Fernández et al., 2012, Beriain et al., 2012,) and temperature with surface acoustics (Cho et al., 2005, Catarinucci et al., 2013, Salmerón et al., 2014). These devices take the form of smart labels, which process the sensor data directly in the RFID tag.

There are also RFID tags and labels that do not require an IC in the transponder, as they encode the data in the frequency or time domain. These are so-called 'chipless' smart labels that can be used in a similar way to conventional RFID chip smart labels. On the basis of these 'chipless' labels, sensing devices for temperature, humidity, gases, visible radiation and surface cracks (Shrestha et al., 2009, Preradovic et al., 2010, Kalansuriya, 2012, Nair et al., 2012, Amin, 2014, Amin, 2014a, Vena et al., 2015) can be developed.

In one study (Escobedo et al., 2016), the authors presented the design, fabrication, characterisation, and validation of a passive printed RFID tag with multiple optical sensing capabilities, which had the ability to measure up to five magnitudes from different sensors with a unique, fully passive printed tag (Figure 5). The tag includes five photodiodes covering a wide spectral range—infrared (IR), visible region (blue (B), green (G), and red ((R) components) and ultraviolet (UV)—and draws power directly from the electromagnetic field that the RFID reader radiates. This type of RFID sensing system provides useful information for detecting spectral data, and its practical applications can be extended to areas such as colourimetry and optical chemical sensing (Capitán-Vallvey et al., 2015).

The tag has been printed on a flexible substrate using screen-printing technology, making it adaptable to different shapes and useful in many smart packaging applications, such as identifying the stage of fruit ripening.

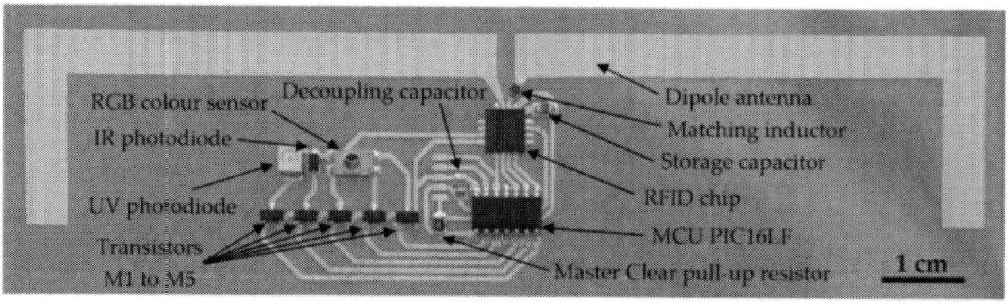

Figure 5. Screen-printed RFID tag on flexible substrate with multiple optical sensing capabilities (Escobedo et al., 2016).

At the present stage of RFID sensing system smart labels, only the RFID tag's antenna and circuit are printed, but the chip and all external sensors are Si-based.

Development of sensing tags is progressing such that besides the antenna, external sensors are also being implemented by using different printing technologies. With reductions in printing cost, new functionalities can be added. The integration of low-cost printed sensors into smart labels opens up a wide ran (Koutsoumanis et al., 2009) of market opportunities to create products with added value and benefits for manufacturers, distributors and consumers alike. Smart labels with printed sensors and RFID or NFC tags can help monitor the conditions a product has been exposed to throughout the supply chain. These smart labels provide real-time data as to a product's temperature, humidity and other data over extended periods of time. If an NFC chip is used in the smart label, data can be read and analysed using a NFC smartphone and printed directly onto the base material; a printed battery can also be used. The thickness of such smart labels is around 1 mm.

1.2. Smart RFID Labels with Printed Sensors

Out of the existing plastic smart labels, multisensors have become more and more popular because, as mentioned earlier, they can monitor different parameters including various gases, humidity, temperature etc. simultaneously. These sensing devices are evaluated in the food logistics chain and integrate passive and semi-active RFID tags using a hybrid approach.

The designs of printed sensors vary and can be printed directly on RFID or NFC tags through the use of various printing technologies (screen printing, inkjet printing, etc.), in a sheet-type application or using roll-to-roll technologies, at low temperatures on plastic substrates.

These systems have begun to be commercialised because of their potential to reduce waste during transportation (Koutsoumanis et al., 2009) Disadvantages of such RFID smart labels are their thickness and high price. Today, the vast majority of commercially available RFID-based sensor systems include only a temperature sensor, although some of them also include a humidity sensor.

In one study (Mraović et al., 2014), the authors presented two types of screen capacitive humidity sensors printed directly on recycled paper and cardboard (Figure 6). Commercially available silver nanoparticle-based conductive ink was used as printing ink. They used one- and both-sided sensors with lateral, comb-type and perforated flat plate capacitor structures. In both cases, the substrate functioned as the dielectric layer.

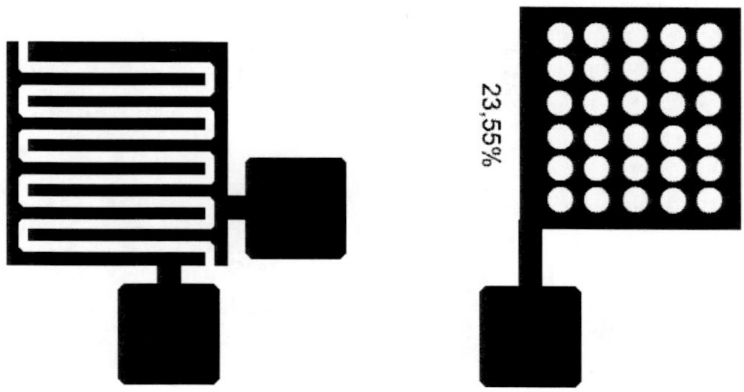

Figure 6. Lateral capacitor (left), perforated flat plate capacitor (right) (Mraović, 2014).

The researchers aimed to define an optimal sensor design (in terms of size and response) for future use as a humidity sensor in paper and board packaging. Analysis showed that such a sensor is in fact the smallest 1 x 1cm comb type sensor, which can probably be made even smaller in future. They concluded that the sensor response is very much more a function of the conductive ink than of the substrate.

A lot of research into prototype models has been developed and demonstrated in various papers recently (Xiao et al., 2007, Lee et al., 2008, Sutija et al., 2013, Sangkil et al., 2013). We can find research works (Mraović et al., 2014, Rivadeneyra et al., 2016) focused on designing, printing and analysing various sensors, including capacitive, resistive, pressure, optical etc. Capacitive structures are widely used in electronics for different applications; these structures are especially interesting in the field of sensors due to their characteristics, such as their low energy consumption, fast response and being non-invasive with no radiation. The most common capacitive sensing structures are parallel plate and the interdigitated electrode.

In one paper (Rivadeneyra et al., 2016), researchers analysed the effectiveness of four planar electrodes developed in only one surface of the substrate, which is particularly useful when the access to a material under test is limited or the other side needs to be exposed to the ambient environment for sensing purposes. Electrodes have been characterised as capacitive sensors, in particular as humidity sensors, printed via inkjet printing using a silver ink on a polyimide flexible substrate, whose electrical permittivity is directly related to the relative humidity.

Capacitive sensors have different geometrical layouts, including interdigitated electrodes, meandered electrodes, spiral electrodes, and serpentine electrodes (Figure 7). Researchers analysed differences between these sensors in terms of fabrication yields, sensitivities to relative humidity, and thermal drift, taking into account frequency dependencies. They proved that all sensors presented similar behaviour in frequency with humidity. Taking into account sensitivity within the same sensing area, the highest value is achieved by serpentine electrodes, followed by spiral electrodes, interdigitated and meandered electrodes. The researchers concluded that the best configuration is dependent on the specific application and its requirements.

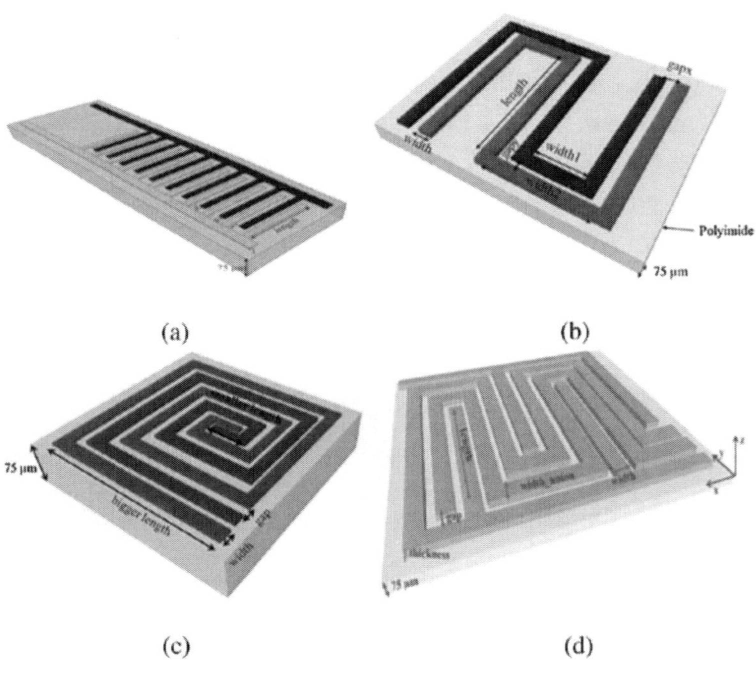

Figure 7. Layout of the designed electrodes: (a) interdigitated electrode IDE, (b) meandered electrode, (c) spiral electrode and (d) serpentine electrode (Rivadeneyra et al., 2016).

A few researchers (Smits et al., 2012, Humbert et al., 2013) have worked on incorporating specific environmental printed sensors together: for example, temperature, humidity and ammonia sensors. The goal is to fabricate fully printed RFID sensing systems that combine a printing-based RFID label and a fully printed multisensor. In recent years, the focus on the development of fully flexible environmental sensors directly printed on polymeric substrates (Briand et al., 2011, Molina-Lopez et al., 2013) has been apparent.

For these sensors, no 'clean rooms' are needed (unlike their Si-based counterparts) because they are processed in a large area using conventional printing methods. While the performance of such sensing systems may be lower, their advantages include lower manufacturing costs, light weight and flexibility.

In the scientific publications, we can find demonstration of feasible printing on polymeric substrates, resistance temperature detectors (RTD) for temperature monitoring (Molina-Lopez et al., 2014), capacitive sensors with cellulose acetate butyrate for humidity sensing (Molina-Lopez et al., 2012), and chemiresistive sensors with conductive polyaniline (PANI)/carbon nanocomposite for ammonia sensing (Wojkiewicz et al., 2011). In one paper, the same research team presented a combination of these sensors on a single polymeric platform and demonstrated its integration onto a printed HF RFID label using techniques compatible with large area manufacturing (Vásquez Quintero et al., 2016).

In another paper (Quintero Vásquez et al., 2016), the authors designed, fabricated and analysed an inkjet-printed multisensing platform (for temperature, humidity and ammonia) (Figure 8) on flexible polymeric substrates integrated into a printed semi-passive high-frequency radio frequency identification (HF RFID) smart label. These sensors were chosen to exemplify the smart label sensing capabilities due to their relevance for many monitoring applications, notably in food logistics chains. The printed platform was integrated after fabrication with the main RFID carrier, which contained an NFC RFID chip, a microprocessor, a frontend readout, and screen-printed circuitry and antenna. The multisensing platform incorporates channels for capacitive vapour (humidity) detection, two channels for resistive-based vapour (ammonia) detection with heating capability, and one resistive channel for temperature detection (RTD).

A modular approach was employed in which the sensing platform was integrated with the base RFID label carrier using foil-to-foil integration techniques compatible with large area fabrication. As well as wireless communication capability, the semi-passive label possesses data logging capability and the capacity to measure data from all sensors simultaneously through direct readout circuitry. The researchers successfully demonstrated the full functionality of the smart label by measuring different temperature, humidity and ammonia levels.

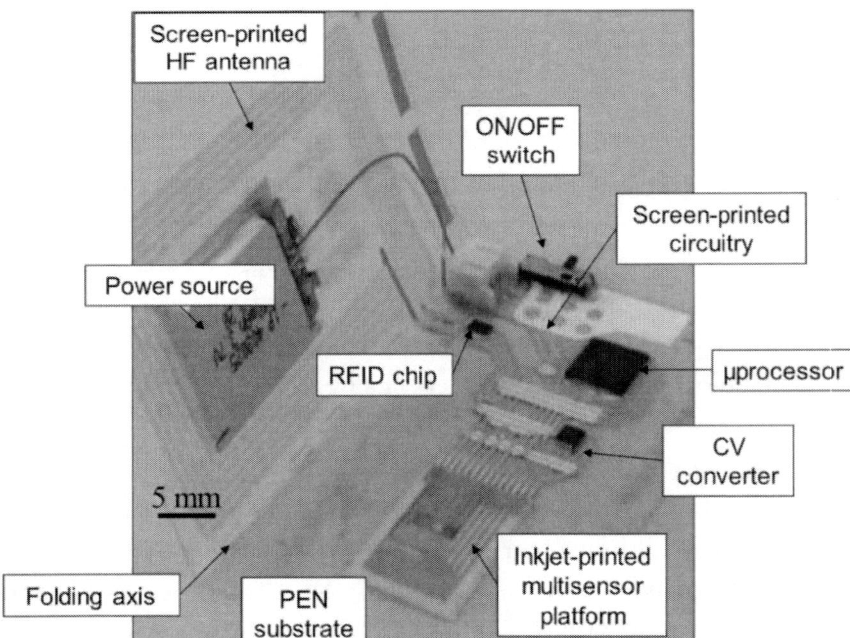

Figure 8. Picture of the final assembly of the RFID half-folded smart label with inkjet-printed multisensor platform (Quintero Vásquez et al., 2016).

1.3. Hybrid Electronics Smart Labels

Smart labels in the near future will need to become thinner and also more flexible in order to prevent physical damage to the labels and enable them to fit more easily in various places on a range of items, such as packaging. Accordingly, it is likely that the next generation of smart labels will solve these problems using flexible hybrid electronics technology (FHE), which combines flexible silicon chips with printed flexible substrates to create products that fit the natural form of packaging (Bringans et al., 2016, Chaney et al., 2017).

Conventional ICs are brittle dies that crack when bent. Flexible electronics have a number of advantages: reduced size and weight, being more rugged and durable, being more environmentally friendly (less material is used), the ability to be placed on surfaces where rigid circuit boards will not fit, and also reduced cost.

The problem is that methods for creating, placing, and interconnecting ultra-thin dies are not commercially available. The methods used for conventional ICs are not suitable. Normally, ICs are 100-300 micron-thick silicon dies that are packaged within a plastic or ceramic enclosure (Figure 9).

Reduction of die thickness below 50 microns means that the dies become flexible, but each of the steps involved (e.g., wafer thinning, wafer dicing, die placement and interconnection) must be re-engineered when wafers and dies become ultra-thin (Uniqarta et al., 2017).

In this way, more complex and larger ICs will enable labels to process the sensor data on the tag. These smarter tags will allow more efficient communications capabilities that will ultimately reduce power requirements and extend battery life. A number of wireless communications will be supported to fit specific application requirements for distance and power, such as < 1 GHz (including RFID and NFC) and 2.4GHz (including BLE) (Bringans et al., 2016, Chaney et al., 2017).

Figure 9. Thin and thick rigid dies on a substrate bent around an item (Uniqarta, 2016).

One possible solution for producing fully flexible smart labels was suggested by Uniqarta (Uniqarta et al., 2017), who developed the ChipStamp™ interposer (Figure 10) for direct connections with circuit board contacts using anisotropic conductive adhesives (ACA). It provides flexibility through the use of a thinned, flexible IC and a flexible interposer material. This results in a low-cost solution for ultra-thin die interconnection. ChipStamp assemblies can be integrated onto printed flexible hybrid electronics devices, like sensing smart labels, using conventional adhesive methods, thus eliminating the need for system integrators to handle ultra-thin ICs.

Flexible electronics will have an impact on many applications such as consumer electronics, healthcare, the automotive industry, pharmaceuticals, asset monitoring, national defence, and also the whole scope of the packaging industry. This will lead us to a new age of ubiquitous electronics with devices embedded in the structures and items around us, in various types of packaging, in our clothing, and even in our bodies.

Figure 10. ChipStamp interposer assembly (Uniqarta, 2015).

2. Materials Used in Smart Labels

The material market for flexible, printed electronics, such as smart labels, is rapidly growing (Vanderroost et al., 2014). The production of smart labels for packaging is emerging and expected to revolutionise this field. A range of substrates and organic and inorganic materials are used in the production of smart labels. Material properties determine the differences between printed and conventional electronics. Printable materials provide decisive advantages in addition to printability, such as mechanical flexibility and functional adjustment by chemical modification.

Inks should be in liquid form as a dispersion, solution or suspension (Bao et al., 2000). In manufacturing processes such as printing, viscosity, solid content and surface tension should be carefully controlled (Weigelt et al., 2010). Interactions (wetting, adhesion etc.) between the layers have an effect on the final properties of the smart label; therefore, the correct materials for use in this process need to be studied and determined.

2.1. Inorganic Materials

For smart labels, it is known that inorganic materials provide highly ordered layers and interfaces compared to organic materials. In general, smart labels provide an additional function in conventional packaging, as they provide knowledge to consumers (Han et al., 2006). In the production of smart labels, the most frequently used inorganic materials for the electrodes of capacitors, inductors, and interconnections are silver nanoparticles, gold particles, copper, aluminium and molybdenum (Chang et al., 2014). Batteries, which should be thin and flexible, are based on zinc and manganese components (Aliaga et al., 2015). Inorganic ink materials are dispersions of metallic or semiconducting micro- and nano-particles (Denneulin et al., 2008, Männl et al., 2013, Willmann et al., 2014). Semiconducting particles include silicon and oxide semiconductors. Silicon is also printed as an organic precursor and converted by pyrolysis and annealing into crystalline silicon (Tanaka et al., 2006, Männl et. al, 2013).

2.2. Organic Materials and Substrates

It is important to understand both the function and the purpose of smart labels made from organic materials, e.g., polymers in general. Organic materials used for smart labels integrate knowledge and developments from electronics, printing and material science (e.g., organic and polymer chemistry) (Vardeny et al., 2005). The basic differences between organic and inorganic materials lie in their structure, operation and functionality, which influences device and circuit design, optimisation and fabrication method (Moliton et al., 2004, Vardeny et al., 2006, Vrabič Brodnjak, 2017). With organic electronics, several layers of different materials are printed onto each other, which reduces the overall number of steps in the manufacturing process and is less costly. At the same time, however, they exhibit worse performance compared to other materials already on the market.

Following the discovery of intrinsically conducting polymers, which first provided organic inks for electronics, the implementation of printed electronics has increased (Moliton et al., 2004). Intrinsically conducting polymers possess semiconducting, conducting, photovoltaic and electroluminescent properties and are not thermoplastic (Brodnjak, 2017).

Semiconductors include the conductive polymers poly(3,4-ethylene dioxitiophene), doped with poly(styrene sulfonate), poly(3,4-ethylenedioxythiophene)-poly(styrenesulfonate) (as PEDOT: PSS) and polyaniline (as PANI) (Sigma-Aldrich, 2017). PEDOT: PSS (Figure 11a) and PANI (Figure 11b) are commercially available polymers and can be used for screen, flexo and gravure printing for smart labels, respectively (Bock et al., 2005).

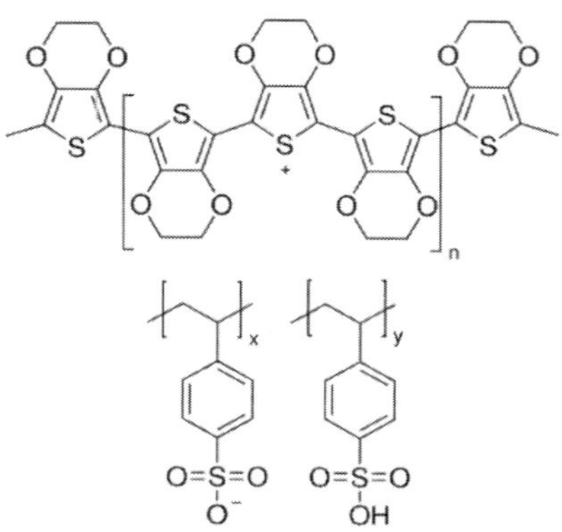

a)

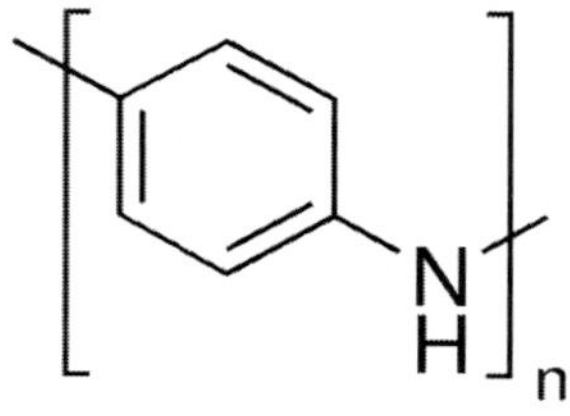

b)

Figure 11. a) Poly(3,4-ethylenedioxythiophene)-poly(styrenesulfonate) (PEDOT: PSS) and b) poly (aniline) (PANI) (Sigma-Aldrich, 2017).

On the other hand, there are other polymers which are also used as insulators and dielectrics. Fibre-reinforced plastic with an outer layer made of silicone rubber or ethylene propylene diene monomer rubber are polymer composite materials used as insulators (Sirringhaus et al., 2000). Such composite materials are less costly and light in weight. However, these materials are still in the research phase, as it has been found that they do not have the long-term proven service life of glass and porcelain (Sirringhaus et al., 2000, Moliton et al., 2004).

In addition, the organic transistor, invented by J.E. Lilienfeld, is a field-effect transistor utilising organic molecules or polymers as the active semiconducting layer (Hasegawa et al., 2009).

2.3. Substrates

The most important substrates for smart labels are silicon, PET (Poly(ethylene teraphtalate)), PEN (Poly(ethylene naphthalate)) and PI (Poly(imide)) foils (Whitmarsh et al., 2005). Flexible polymer films are still typically used as support or base material (Moliton et al., 2004, Perelaer et al., 2009).

Most of the traditional electronic components used today in RFID manufacturing are still silicon-based (Chaves et al., 2010). These electronic components are high-performance, highly miniaturised and integrated and their price is low. On the other hand, the production of smart labels from silicon requires costly clean room facilities and the manufacture is complicated (Chaves et al., 2010). The basic design of an organic smart label is virtually no different from a regular silicon-based RFID (Chaves et al., 2010, Willmann, 2014). Currently, research is being conducted into how optical and electrical circuits can be integrated in organic materials, using polymers instead of silicon (Vanderroost et al., 2014).

Poly(ethylene teraphtalate) foil is a common choice, due to its low cost and high temperature stability, while poly(ethylene naphthalate) and poly(imide) foils are higher-performance and higher-cost alternatives (Whitmarsh et al., 2005, Denneulin et al., 2008, Chaves et al., 2010). Paper, board and cardboard are attractive substrates, but are problematic for smart labels due to their high roughness, water and moisture absorbency.

Smart labels produced on flexible substrates lower production costs and allow for easier fabrication (Vanderroost et al., 2014). Although there have been many studies conducted on paper, its high roughness, high moisture absorbency, low durability and worse mechanical properties compared to the abovementioned materials make it problematic for use in electronics (Denneulin et al., 2008, Denneulin et al., 2011, Willmann et al., 2014). Consequently, applications such as special coatings and Corona discharge could improve the properties of paper discussed above. Special additives, which are often used in conventional printing inks, are not appropriate in this context, since they defeat electronic functionality (Chang et al., 2014, Willmann et al., 2014). On the other hand, there are organic electronics created from polymers, which are used as insulators, dielectrics, sensors etc. Compared to silicon, polymer electronics offer the advantage that they can be directly fabricated onto flexible substrates by printing technologies (Bock et al., 2005, Sigma-Aldrich, 2017).

2.4. Future Trends in Materials for Smart Labels

There are many good ideas based on these applications, and consequently a need for new materials exists. The flexibilities of the substrates, conductivities of the inks and stability of polymer resistors need to be increased so that they can compete with copper in smart label applications. In the future, organic materials in smart labels are expected to replace traditional silicon-based electronics in cases where smart labels do not need to be very fast, but should be small and low-cost.

3. Technologies

Printing technologies as deposition techniques have long been a part of printed circuit board manufacturing (PCB). However, printed electronics—especially smart and flexible label technologies—have created new issues that demand to be solved. Almost all printing technologies involve additive patterning (a process where the material/ink is applied or transferred to a substrate), while more conventional PCB manufacturing mainly uses subtractive patterning (spin coating, vapour and electrochemical deposition, etc.) with multiple production steps, as presented in Figure 12.

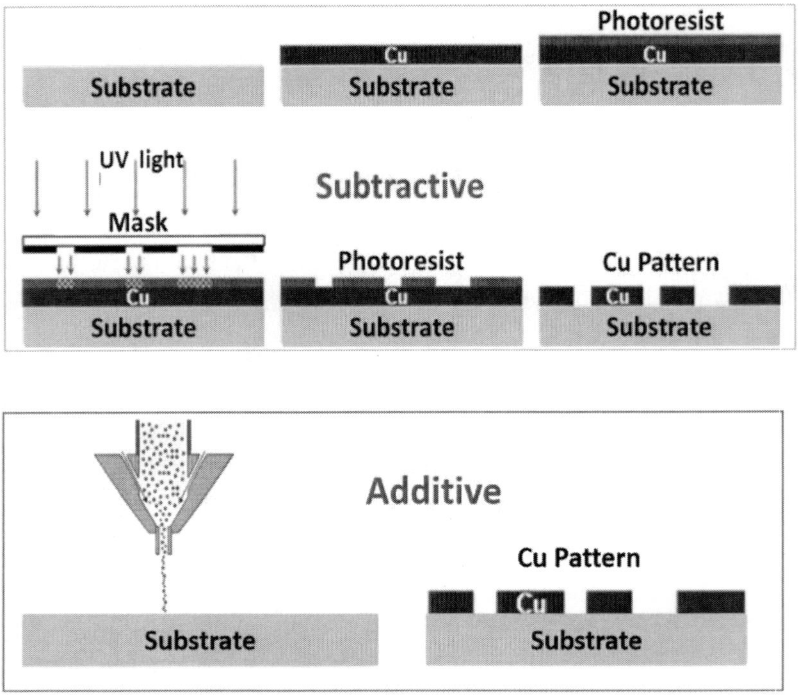

Figure 12. Additive vs. Subtractive manufacturing process.

Table 1. Comparison of traditional (classic) printing methods and electronics printing requirements

Requirements	Traditional printing	Electronics printing
Resolution	15-100 μm	<< 20 μm
Registration	Low	High
Edge sharpness	High	Very high
Uniformity of layers	Low importance	Very important
Adhesion of layers to substrate	Important	Important
Adhesion of layers to other layers	Important	Very important
Solvents in inks	Cost issues	Functional issues
Purity of solution	Not very important	Very important
Visual properties	Very important	Not important
Electrical properties	Not important	Very important

This is one reason why classic printing technologies are attractive for printed electronics applications, in spite of their inherent limitations. As the requirements for printed electronics differ from the expected classic printing products (Table 1), the appropriate technologies should be carefully chosen on the basis of the production requirements.

One of the printing technologies already used for smart labels and other printed electronics is screen printing.

The conductive ink depositions on rigid boards are printed in special control rooms using screen printing, which is mainly employed due to the high range of ink deposition values in microns, the fairly low cost of printing large and non-standard-sized products, and the easy transfer of conductive paste inks. While other printing technologies were mainly designed for paper and print media and packaging, screen printing was always more of a product-oriented ink deposition technology. On the other hand, it exhibits some shortcomings regarding speed and accuracy in the roll-to-roll application, which is the most productive regarding deposition speeds. However, flexo printing (direct printing with printing forme), gravure (direct printing with printing forme) and offset printing (indirect printing with printing forme) are established technologies for classic (not smart) labels, and packaging and usability with these technologies would be price-efficient, as they allow for one-pass printing. The emergence of digital printing technologies enabled very precise ink deposition on a very wide range of substrates/carriers, but these are still lagging when compared to rotary or web-operated classic printing technologies in terms of speed and formats. Almost all existing printing technologies have already been used and evaluated for printed electronics, including gravure (Higgins SG et al., 2015, Voigt et al., 2010, De La Fuente Vornbrock, 2010), screen (Peng et al., 2013, Chang et al., 2009,), inkjet (Vásquez Quintero et al., 2016) and flexo (Sanha et al., 2016), as well as fine pattern-printing technologies such as nanoimprint (Kumar et al., 1994), μCP (Yin et al., 2010) and electrostatic inkjet. These last three technologies are so rare that they can only be used for very specific printing applications and are not contenders for high-volume smart label printing. In classic printing, there are so-called 'hybrid' printing presses where the colour component is done (for example) in offset and additional surface enhancement is achieved via flexo or inkjet; however, this combination of technologies is currently far from being developed for the PE market. A short overview of printing technologies is presented in Table 2.

Table 2. Overview of printing technologies

Printing method	Printing forme	Direct/ Indirect	Ink viscosity (cP)	Line width (μm)	Line thickness (μm)	Printing speed (m/min)
Offset	Yes	Indirect	100-10,000	Approx. 19	Several 10	Sheetfed middle Web offset fast 1000
Gravure	Yes	Direct	100-1000	10-50	Approx. 1	Fast 1000
Flexo	Yes	Direct	50-500	45-100	<1	Fast 1000
Screen	Yes	Direct	500-5000	30-50	5-100	Middle 70
Inkjet	-	Direct	10-20 (electrostatic approx. 1000)	30-50 (electrostatic approx. 1)	1	Slow to Middle 10
Dispense	-	Direct	$1000\text{-}10^6$	Approx. 10	50-100	Middle
μCP	-	Direct	-	Approx. 0.1	Approx. 1	Slow
Nano-imprint	-	Direct	-	Approx. 0.01	Approx. 0.1	Slow

The main challenges for all printing and deposition technologies are registration (alignment of fine patterning), chemical compatibility of materials, physical compatibility (coverage in the appropriate amount and formation), and curing drying challenges in the prescribed time/temperature limit range. For example, offset, which has a good registration and reasonably fast production due to the use of dampening solution for making printing and non-printing areas, as well as highly viscous inks which can induce picking due to high forces in the printing NIP, is almost unsuitable for controlled printed electronics production. On the other hand, for example, gravure, which is used in high-speed, high-quality large print runs in the printing industry, has some shortcomings, such as the high price of printing forme, the printing pressure applied by the gravure roll during the ink transfer to the substrate (which can damage soft materials and very sensitive connections) and the inability of the doctor blade to effectively wipe the non-required ink from the non-printing areas, which can cause problems in the electric layer. The thickness of the printed layer generally determines the electrical conductivity. This thickness will largely depend on the printing process, printing ink rheology and substrate absorbing capacity. In order to increase electrical conductivity, the printing can be modified so that a thicker ink layers are deposited, or by simply increasing the number of printing passes. However, this will all depend on the behaviour and capability of the printing substrate and the number of available print stations. Many applications also require a lower thickness of the deposited ink film. This can pose problems in obtaining the desired conductivity of an ink film or the printing of a thin uniform dielectric layer and more than one deposition technology is needed. Out of all the classic printing technologies, inkjet and screen printing have promising features regarding printed smart labels. Printed smart labels and other flexible materials have their challenges regarding manufacturing steps due to the flexibility/rigidity of the materials and process handling. Pattern processes for flexible electronics should ideally allow: noncontact patterning, low-temperature process, additive patterning, real-time adjustment, three-dimensional structuring, easy registration, printability of organic/inorganic materials, low cost, and compatibility with large-area and high-throughput processing (Zhang, 2014). Challenges of inkjet printing are mainly concentrated around ink formulation (viscosity, surface tension and evaporation rate), substrate quality (surface energy), jetting reliability, ejector variations and print resolution. The inkjet technology is broadly classified into two categories, based on the mechanism of droplet generation: continuous inkjet (CIJ) and drop-on-demand inkjet (DOD). DOD printing is, in turn, classified into three types: thermal inkjet, piezo inkjet and electrostatic inkjet.

An overview of the typical parameters of inkjet solutions is presented in Table 3 (Zhang, 2014).

Table 3. Typical parameters of inkjet solutions

	Piezo DOD	Thermal DOD	E-Spinning	E-Spraying	EHD Jetting
Viscosity (mPa/s)	<40	1.28	7.175	2.76	-90
Surface tension (mN/m)	20-70	Temperature dependent	15-64	-45	-50

One of the typical inkjet printing problems is solving the Coffee Stain Effect, which is a result of the uneven evaporation of a liquid over an inkjet ink droplet liquid air interface. Some work has been done using (for example) Aerosol Jet printing, which solves some of the aforementioned difficulties in inkjet printing. Aerosol technology allows easy printing of 3D objects with a wide range of viscosities and particle sizes ranging from nano to micro. The sintering is done via laser, which enables low curing and melting temperatures.

Inkjet technology has a good predicted future for all printed electronics applications, but speed and the economic viability of printing is required to achieve full-scale production. Other limiting factors are the ink development of Cu-inks in nanoparticles as well as the thermal sintering alternative.

On the other hand, screen printing, which currently dominates the market over other printing technologies, has a wide variation of influencing factors such as frame registration, stencil and mesh quality, mesh resolution and deposited ink film thickness, squeegee material speed, and angle, which also affect deposition quality and surface energy interactions. The market gain is due to its being a well-established process with a reasonable price and adequate development of conductive inks. With its fairly good registration, line thickness and stability, screen printing will remain a dominant printing technology for some years to come.

4. APPLICATIONS

According to Statistics MRC, the global smart packaging market, which includes intelligent, active, and modified atmosphere systems, had an estimated value of $15.7 billion in 2015 and is expected to reach $32.7 billion by 2022 (Market research consulting, 2016). The market share of intelligent packaging is below 15% at time of writing, but is expected to grow at an estimated CAGR of over 12% over the forecast period (Grand View Research, 2014). Intelligent packaging is expected to be the fastest-growing segment of smart packaging, mostly because of the increasing use of indicators, especially time-temperature indicators, and the product differentiation, traceability and other interactive features offered at more affordable prices by intelligent systems (Lee and Mijanur Rahman, 2014). Smart labels, as part of intelligent systems, will reach $10.03 billion by 2021, growing at an estimated CAGR of 17.65% (marketsandmarkets.com, 2014).

There are many reasons to label items, including logistics, supply chain management, improving efficiency of the fabrication control process, complete product traceability, product liability, quality guarantees, and recyclability issues (Weigelt et al., 2010). In particular, traceability, ensuring product quality (especially food quality) and safety are of major concern in today's society (Bibi et al., 2017). Therefore, identification technologies—in particular RFID systems—are becoming increasingly important.

Because of the increasing demand for labelling, smart labels must be produced in very high volumes, and the production process should be as simple and cost-efficient as possible. Printed electronics technology has the potential to meet these demands. According to Kleper (Kleper, 2004), one of the first market applications of printed electronics was RFIDs embedded in cartons, besides passive ID cards printed on paper used for ticketing or toys (Hecker and Breitung, 2013). The market for printed electronic devices is constantly growing. Estimated at over $2 billion in 2011, and over $24 billion in 2014, the global market is expected to increase to $45 billion in 2021 (Kirchmeyer et al., 2015). Out of the different devices, displays and organic solar cells are predicted to take the lead by 2020, while smart devices (e.g., RFID tags, sensors, indicators) are expected to expand their market later (Park, 2013). The benefits of printed electronic devices are their mechanical flexibility, thin form and light weight, meaning they can be integrated as smart labels in packaging and sensor networks (Street et al., 2015). Applications include RFID tags and sensing devices.

4.1. RFID Devices

RFID devices are one of the technologies used for automated data collection, as well as being one of the major technologies in the field of identification. Compared to RFIDs, barcodes have limited capabilities; however, because they are cheaper, they are still prevailing on the market. In spite of this, RFID systems are gaining more and more popularity in numerous sectors (Kumari et al., 2015, Realini and Marcos, 2014, Xiao et al., 2007). According to Bibi et al., more than 3,000 cases in different domains of applications are known (Bibi et al., 2017). Current RFID applications, according to Gaukler and Seifert, are in 1) retail; 2) logistics, transportation and warehousing; 3) assembly, manufacturing and configuration management; 4) asset tracking and locating objects; 5) authentication, counterfeit protection and security (Gaukler and Seifert, 2007). Some of the applications of printed RFID labels include airline baggage tracking, ID and tracking for express parcels, product ID and tracking in the supply chain, brand authentication, document tracking, and library material handling. Tagging in the supply chain includes Consumer Units (products and individual items), Traded Units (boxes/packaging/product carriers), and Distribution Units (pallets/trucks) (D'Hont, 2004).

Passive RFID tags are typically used in product tracking, including applications in access control, identification, location and security, whereas active and semi-passive tags are used for higher-value goods that are scanned over longer distances (Angeles, 2005). Walmart was one of the first large companies to use RFID in retail, demanding that vendors equip merchandise shipments with RFID tags and to test piloting smart shelves (Kelly and Erickson, 2005). Similarly, more than ten years ago, the German company Metro introduced RFID on the pallet level, as well as on a case-level basis, throughout its supply chain. Item-level RFID tags were used on only four products displayed on smart shelves at the Metro Future Store (Gaukler and Seifert, 2007). Other companies followed and began to use passive RFID tags in the management of their supply chain, mainly on the case/pallet level, in order to improve backroom inventory-management practices (Zhu et al., 2012). Subsequently, stakeholders in the apparel industry such as Levi Strauss & Co., Marks and Spencer and Zara have also begun to use RFID tags on the item level (Ilie-Zudor, 2011). A product with an RFID tag is tracked immediately after it is delivered to the store. The use of RFID in smart shelves and on individual items improves data accuracy, information visibility and store security; it can also assist in minimising backroom inventory, reducing inventory shrinkage, keeping shelves full and preventing both stockouts and excess stock. The CPG (consumer packaged goods) industry could benefit a lot from such RFID inventory monitoring by converting it into FMCG (fast moving consumer goods) (Zhu et al., 2012).

RFID technology can be used to create end-to-end supply chain information visibility by enabling suppliers, manufacturers, logistics providers and retailers to track and trace item-level information throughout the entire supply chain at any time and at any location (Zhu et al., 2012). Delen et al. showed that information collected via RFID can be valuable to both the retailer and the supplier (Delen et al., 2007). In transportation, an RFID system is used for logistics tracking, delivery security, monitoring, and toll collection, and can also be used as a smart key or in public transport and ticketing (Ilie-Zudor et al., 2011). At the warehousing level, RFIDs can help with receiving, handling and shipping management of stored products. Unilever and Coca Cola Germany are two companies that apply an RFID-equipped smart pallet system designed to move, handle and track products in warehouses (Angeles, 2005, Ilie-Zudor et al., 2011). Especially for perishables, RFID can implement an infrastructure that enables more sophisticated control and handling of goods, for example keeping products within safety regulations and time-dependent dispatching.

In the agriculture and food sector, the traceability of products across the whole supply chain (production, transportation, storage and delivery) is very important in order to ensure food quality and safety. Effective tracking, meaning the ability to locate a food product or ingredient and follow its history in the supply chain both forward (from the source to the consumer) and backward (from the consumer to the source) is becoming an important issue for all parties, including the consumer (Costa et al., 2012). It is not surprising that the first cases of RFID use began more than ten years ago. Mennecke and Townsend examined the feasibility of extensive RFID tagging to determine product provenance in the meat production industry (Mennecke and Townsend, 2005). E-Provenance has launched RFID to monitor the quality of fine wines and trace their origin (Lee and Mijanur Rahman, 2014). RFID tags linked to a web platform have been used for vegetable supply chain traceability (Hu et al., 2013), a traceability system for animal product safety (Ben Hai et al., 2007), and monitoring the production history and quality of cheese (Papetti et al., 2012).

4.2. Sensing Devices

Indicators are devices that indicate the presence, absence, or concentration of another substance, or the degree of reaction between two or more substances, by means of a characteristic change (thefreedictionary.com, 2017). They provide immediate qualitative information through direct visual change, which must be simple, continuous and understandable for consumers in order to be effective (Kerry et al., 2006). They can indicate elapsed time, change in temperature, humidity, concentration of a defined substance, or shock abuse (Rodrigues and Han, 2003).

Among the intelligent systems, temperature indicators, which are usually applied in the form of labels on packaging, have the most commercial applications. Fluctuations in temperature may cause deterioration in quality and safety for many products, such as chilled and frozen food, as well as certain medical and pharmaceutical products (Lee and Mijanur Rahman, 2014). Temperature indicators, sometimes called critical temperature indicators (CTI), are simple thermal recorders which show whether products have been heated above or cooled below a certain temperature (Pault, 1995). In this case, only the critical temperature is displayed without other information. In addition to showing whether the appropriate temperature has been reached for food or beverage to be consumed, CTI can indicate denaturation of an important protein above the critical temperature or the growth of a pathogenic microorganism (Pavelkova, 2003). CTI mostly uses thermochromic ink to indicate if the product is at the correct serving temperature. Company Sun Chemical developed a reversible thermochromic ink on pizza packaging to indicate that the product is hot enough to eat; if the pizza cools down, the label loses colour (Smithers Pira, 2013). Hallcrest OK® (LCR Hallcrest Ltd) labels for fridges are used to indicate when the temperature is out of range (LCRHallcrest, 2016). For beverages (such as beer, alcoholic drinks and soft drinks), CTI labels show when the drinks are chilled and ready to be consumed, whereas those used for pizza and ready-meals show whether the product is hot enough. Usages for food safety include revealing when a required temperature has been achieved, indicating the temperature range at which it is proper to keep fruits, milk, or other food products, showing when bread is baked, meat is properly cooked or an egg is boiled, and alerting consumers or staff when food is out of date (Smithers Pira, 2013). They can also be used for promotional purposes and marketing.

Time-temperature indicators (TTI) show the accumulated time-temperature history of a product and give a continuous, temperature-dependent response throughout a product's history (Robertson, 1993). They provide visual information about temperature history during distribution and storage and indicate whether the package has been improperly exposed to high temperature, as well as the effect of this exposure, based on its duration (Mohebi and Marquez, 2015). The operating principle is grounded in mechanical, chemical, electrochemical, enzymatic or microbiological change, usually expressed as a visible response in the form of a mechanical deformation, colour development or colour movement (Taoukis and Labuza, 2003). TTIs are useful in indicating breakdowns in the distribution chain and for products in which reactions important to quality or safety are initiated or occur at measurable rates above a critical temperature. Examples of such reactions are microbial growth or enzymatic activity that is inhibited below the critical temperature (Pavelkova, 2003).

The performance of TTIs has been evaluated with respect to a range of different food products, such as chilled fish, dairy products, frozen beef, frozen vegetables, tomatoes and lettuce, and canned fruitcake (Lee and Mijanur Rahman, 2014). Though there are already several commercial applications of TTIs, such as 3M™ Monitor Mark™, Fresh-Check®, OnVu™, CheckPoint® and TopCyro®, the development of TTIs is an object of continuous research. 3M™ Monitor Mark™ (3M Company) TTI label, which is based on the diffusion of blue-dyed fatty acid ester at a temperature-dependent rate, monitors product exposure to time-temperature history (Pavelkova, 2003, Gregor-Svetec, 2017). The Fresh-Check® (Temptime Corporation) TTI label is based on the polymerization of diacetylene and matches the shelf life of the food products to which it is affixed. A part of the label darkens if the package has experienced unfavourable temperature exposures, and the intensity of the colour can be easily compared to the reference colour scale on the label (Pavelkova, 2003). OnVu™ (Freshpoint/BASF) is based on the photochemical reaction of temperature-sensitive ink that changes colour with time at rates determined by temperature. TTI labels show a temperature memory, for example for chilled and frozen products (Pavelkova, 2003, Gregor-Svetec, 2017). The CheckPoint® (Vitsab International AB) TTI is based on a pH-dependent enzymatic reaction that triggers a colour change. This label reacts to time and temperature in the same way as the product, and thus provides information about its state of freshness and remaining shelf-life. It is used to monitor storage environments for time-temperature abuse and food safety (Pavelkova, 2003, Gregor-Svetec, 2017). TopCyro® (Crylog) is a microbiological TTI that monitors the time-temperature history of the cold chain in real time.

Gas indicators (GI) can monitor the changes in gas composition inside the package, and thus the safety and quality of products, or work as leak indicators giving information on package integrity throughout the whole distribution chain (Gregor-Svetec, 2017). They change their colour due to chemical or enzymatic reactions, typically providing information about the presence or absence of oxygen or carbon dioxide (De Jong et al., 2005). The main application of oxygen indicators is in combination with the oxygen scavenging system. One commercial indicator, AgelessEye® (Mitsubishi Gas Chemical Company, Inc.), is a reversible oxygen indicator based on the oxidation/reduction of redox dye. It works as a leak indicator giving information on package integrity, mostly for modified atmosphere food packaging (Bibi et al., 2017). Carbon dioxide indicators based on pH-dependent colour changes of dye provide information about the concentration of carbon dioxide in the packaging, and also detect early spoilage. As the concentration of carbon dioxide changes, it alters the pH, which changes the colour of the indicator (Hong and Park, 2000).

Freshness indicators (FI) are sensors that monitor the quality of packed food by reacting to the changes taking place in food as a result of chemical changes, microbial growth or metabolism due to exposure to detrimental conditions or exceeded shelf life (Ghaani et al., 2016). Indicators are based on an indirect detection of metabolites through colour indicators (e.g., pH) or a direct detection of target metabolites using biosensors (Realini and Marcos, 2014). They provide direct product quality information instead of merely indicating temperature abuse or package leaks (Smolander, 2008). RipeSense™ (Ripesense, 2010) is the world's first intelligent label that colourimetrically indicates the degree of preferred ripeness for fruits based on detecting ethylene, which is released by the fruit in the ripening process (Pocas et al., 2008). SensorQ™ (Food Quality Sensor International Inc.) is a pH-sensing colourimetric indicator of microbiological origin for packed meat and poultry. Other freshness indicators are based on sensitivity toward other microbial metabolites, such as ethanol, diacetyl, and carbon dioxide (Pereira de Abreu et al., 2011).

Sensors are devices which are used to detect, locate or quantify energy or matter, giving a signal for the detection or measurement of a physical or chemical property to which the device responds (Kress-Rogers, 1998). Gas sensors are devices that respond reversibly and quantitatively to the presence of a gaseous analyte by changing the physical parameters of the sensor and are monitored by an external device (Kerry et al., 2006). OxyDot-O2xyDot® (OxySense Inc.) is the first commercially available fluorescence-quenching sensor. Among others, it has been used to determine the amount of oxygen present in sealed food (Lee and Mijanur Rahman, 2014).

Biosensors detect, record and transmit information pertaining to biological reaction (Yam et al., 2005). They are able to monitor food freshness in a more specific way than freshness indicators, as they can detect the formation of degradation products and might be designed in a tailored manner according to the type of product being packed. Biosensors can also detect allergen proteins in foods to prevent reactions to peanuts, tree nuts, and gluten (Kuswandi et al., 2011). Food Sentinel System™ (SIRA Technologies) is a biosensor developed to detect food pathogens with a specific-pathogen antibody, that is attached to a membrane forming part of a barcode. The presence of contaminating bacteria causes the formation of a localized dark bar, rendering the barcode unreadable upon scanning (Yam et al., 2005).

Colourimetric sensors enabling the "real-time" monitoring of spoilage offer the potential of developing dynamic "best-before" dates and can lead to improvements in quality assurance (Kuswandi et al., 2011).

4.3. Sensing Systems Interfaced with RFID Tags

More evolved RFID systems allow the integration of other functions into the RFID tag, such as time-temperature indicators or biosensors, to monitor and communicate the temperature history of the product, as well as to give quality information by obtaining microbial data (Guillory and Standhart, 2012). By evaluating the environmental conditions, the quality of a product and its shelf life can be estimated, and decisions for optimising logistics can be supported (Bartels et al., 2010). Sensors may be embedded as passive, semi-passive and active RFID tags, either directly onto the antenna or connected to the chip, with logging features to keep track of measured quantities, either as a function of time or by recording predefined events (Ruhanen et al., 2008). The coupling of sensors, such as time temperature indicators (TTI), biosensors, gas sensors and humidity sensors, to RFID not only improves monitoring of product quality and management of the supply chain, but also enables the creation of a web database available to suppliers and consumers (Bibi et al., 2017). The advantages of RFID coupled with TTI are numerous: lower costs (TTI-RFID tags are cheaper than temperature loggers); more precision than traditional temperature measuring systems; the ability to record the history of the cold chain, and the ability to provide real-time information with the coupling of the system to GPS and mobile networks (Atsushi and Kentaro, 2006, Kumari et al., 2015). Temperature sensors embedded with RFID tags were used for cold chain monitoring of frozen foods and temperature monitoring of seafood products, dairy products, beverages and horticultural products (Hu et al., 2013).

Humidity sensors, which perform relative humidity monitoring together with RFID tags, could be used in the agricultural and food sectors, the medical and pharmaceutical sectors, and sectors dealing with goods that are perishable or vulnerable to environmental conditions. Abad et al. used a humidity sensor connected to the chip of a smart tag for real-time traceability and cold chain monitoring for seafood products packed in polystyrene boxes (Abad et al., 2009). The wireless pH sensing system, which included a flexible pH sensor, was tested for *in situ* monitoring of the spoilage processes in fish products (Huang et al., 2012). Sen et al. estimated the freshness of pork using a monitoring system consisting of an RFID tag and a temperature and gas sensor (Sen et al., 2013). RFID tags with sensors able to measure temperature, humidity and the presence of volatile amine compounds were used by Smits et al. to estimate the freshness of cold fish (Smits et al., 2012). Electrochemical detection of volatile chemicals has also been developed, working on the resistance changes of RFID tag circuitry, including changes in the frequency and amplitude of the output signal. Other smart tags with sensors developed for potentiometric and conductometric responses, optoelectronic interface and photometer based have also been studied (Bibi et al., 2017). To date, only a few commercial applications of RFID tags with sensing functionality are known, mostly with integrated time-temperature indicators (TempTRIP® sensor tags, Easy2log©). It is reported that use of Easy2log© or TempTRIP® tags, with the combined use of RFID and indicator in the cold chain, results in the improvement of supply chain management efficiency (Realini and Marcos, 2014).

Jedermann et al. showed how RFID temperature loggers can be adapted to analyse the amount of local deviations, detect temperature gradients, and estimate the minimum number of sensors that are necessary for monitoring of the refrigerated transport of perishable goods (Jedermann et al., 2009). In a study by Badia-Melis et al., refrigerated fruit storage monitoring was performed combining two different wireless sensing technologies: RFID and wireless sensor networks (WSN) (Badia-Melis et al., 2015). The future applications of RFID tags equipped with sensors are in integration with WSN towards pervasive systems for long-term and large-space monitoring. WSN can operate in different environments and provide longer reading range as well as other benefits, such as different network topologies, flexibility, reliability, accuracy and a variety of sensors that are already implemented (Rawat et al., 2014). The tags and sensors on products would be able to communicate and share data with each other, thus improving the efficiency and effectiveness of product-condition monitoring and management (Cain and Lee, 2008). Due to their low cost, ease of use and convenient integration, RFIDs coupled with sensors can play an important role in pervasive IoT applications in the future (Meng and Li, 2016).

5. Environmental Impact of Smart Labels

Of the existing types of smart labels, RFID tags have the largest market share. The total RFID market is currently worth \$11.2 billion in 2017 and is forecasted to rise to \$14.9 billion in 2022 (Das, 2017). Most of that growth is due to passive RFID tags. According to IDTechEx, in 2010 the market share of active tags was less than 3 percent, with a total of approximately 60 million tags, compared to 2,250 billion passive tags (Schindler et al., 2012). The application area with by far the largest number of tags in the future will be consumer goods, with very strong growth predicted in this area starting from 2018. Given the increasing demand for RFID tags, their impact on the environment will become an important issue in the future.

5.1. Negative Impact

Disposal of smart labels, like that of any other items, must follow the waste hierarchy and its priority order outlined in the Waste Framework Directive (Directive 2008/98/EC, 2008). Smart labels and active RFID tags are categorised as electronic devices and therefore fall under the scope of the WEEE Directive (Directive 2012/19/EU, 2012). However, smart labels are mostly attached to packaging, meaning that the Packaging Directive must also be considered (Directive 94/62/EC, 1994). Moreover, passive RFID tags are not categorised as electronic devices; most of these are in the form of adhesive labels attached to consumer packaged goods (CPGs), made out of cardboard, paper and plastics, but also metal, glass, wood, etc. Because they are attached to other objects, the end-of-life phase for RFID tags depends on the item they are attached to. Sorting of materials into recycling streams depends on material, treatment, complexity and hazardousness (Schindler et al., 2012). In this way, RFID tags and their components enter into different waste treatment paths.

Disposed items that fall under the Packaging Directive are required to be recovered and recycled (Directive 94/62/EC, 1994). Energy recovery is performed through thermal conversion processes, where both the energy content of the material that is combusted and the gaseous components of the incineration process (that could increase emissions) are important issues (Schindler et al., 2012). The impacts of RFID tags are not expected to be of any importance for the process steps. However, there may be some impacts resulting from the treatment of the incineration output streams due to the increased presence of heavy metals, in particular copper, in bottom ash and gaseous output, where cleaning must ensure that emission limits are not exceeded (Schindler et al., 2012).

Most RFID tags will enter material recycling paths. The metallic components coming into consideration for material recycling of RFID tags are copper, silver, aluminium and silicon. The recycling of RFID tags via the copper metallurgical process was suggested by Schnideritsch et al. due to its collector capacity for precious and other metals (Schnideritsch et al., 2012). Until now, selective separation of RFID tags from other waste components has not been feasible, and neither has the allocation of RFID components to specific recycling paths (Schindler et al., 2012). Most RFID tags will enter the recycling process of the items to which they are attached. In the case of material recycling, the purity of the recyclate is very important. Even if RFID tags have no impact on the recycling process, they do have an impact on the processing costs and/or the quality of products, material loss and the recovery rate (Schindler et al., 2012).

In the recycling of metals, alloying elements expected from RFID tags are copper, silver, nickel and silicon. These can have positive or negative effects depending on the alloy produced (Kräuchi et al., 2005), In recycling of ferrous materials, copper and silver are unwanted elements, whereas aluminium and silicon have no impact on the quality of steel, though they do increase the amount of slag (Schindler et al., 2012). Assuming a copper concentration of 0.1% coming from one RFID tag-equipped steel can, the copper content in steel would build up and eventually reduce the quality of steel if a larger number of cans are recycled (Thomas, 2008).

Recycling of non-ferrous metals is accomplished via the copper and aluminium route. RFID tags might lead to an accumulation of copper in aluminium recycling in the long term, lowering the quality of aluminium, although they have no impact on the copper route (Schindler et al., 2012). The situation varies depending on the intended application of the recycled aluminium. An amount of impurities of a magnitude of 0.1% to 5%, with copper up to 4%, is allowed in some cases (Kräuchi et al., 2005). For high quality products, however, no contaminants are allowed in aluminium recycling, and contaminants must be screened out or removed via other means (Thomas, 2008). Accordingly, recycling RFID tags via the copper route is considered the most promising way of recovering both the energy content as well as most metals (especially the precious metals) contained in RFID tags (Schindler et al., 2012).

The non-combustible components of RFID tags, such as silica can impact the glass recycling process and the quality of the glass if they are not removed before smelting. Silicon melts at a different rate to glass, resulting in "silicon balls," which can cause potential weak spots in a new glass product (Thomas, 2008). Copper included in the smelting process could cause discolouration, whereas inclusion of aluminium could lead to material defects in a produced glass item (Kräuchi et al., 2005). RFID tags used on glass products therefore need to be completely removable or placed on the caps rather than on the glass itself (Schindler et al., 2012).

Most components of RFID tags, i.e., the electronic chip, metal in the antenna, polymer-based substrates, or solder used to join the antenna and chip, could also interfere with the recycling process of plastic materials and lower the quality of the recycled product (Schindler et al., 2012). Inclusion of any metal in any kind of plastic recycling is problematic, as unfused metals cause discontinuities in the recycled plastic product (Aliaga et al., 2011). Accordingly, RFID tags used on plastic products should be removable in the cleaning process. The shredding process is also problematic because the risk of damaging the tag is high, resulting in small shredded pieces which will enter the recycling process as impurities. The separation of RFID tags takes place in the density separation step. Since RFID tags have higher densities than some polymers, like polyolefins, they can be effectively separated in water. However, some amount of metals and other impurities is expected to pass the separation step and enter the extrusion process, where they can create disturbances such as blocking the screens or nozzles of extruders (Schindler et al., 2012). Moreover, due to impurities present in the melted polymer, the quality of recyclate is also affected. It was shown in a study by Aliaga et al. that the quality of recycled HDPE was not lowered due to the presence of RFID tags, although some processing problems were observed (Aliaga et al., 2011). For RFID tags attached to PET or other polymer of similar density, separation based on mass difference is not appropriate, and consequently a number of problems are expected in the recycling process.

In the recycling process of paper and cardboard, RFID tags are separated in the pulping process; for paper-based RFID tags, the metals and the chip remain as solid objects, whereas plastic labels remain as complete objects (Schindler et al., 2012). The items that are not dissolved are then subjected to density separation processes. In the deinking process, if applied, components from RFID tags are further removed. It is expected that not all the materials will be completely separated. However, according to some studies (Schindler et al., 2012, Furuta et al., 2008), no direct problems for the technical processes of paper recycling are expected. On the other hand, RFID tags can influence the composition and quantity of solid and liquid residues. Adhesives, along with very small plastic and metal parts, will pass through the filter system, consequently increasing the number of stickies in recycled fibres and degrading paper quality (Schindler et al., 2012, Kräuchi et al., 2005).

Pilot scale recycling, based on the industrial process of pulping, screening and paper formation, was performed by Aliaga et al. on envelopes containing printed antennas, silicon-based resistors and batteries (Aliaga et al., 2015). According to the results they obtained, resistors were removed by the screens as reject, whereas the battery and nano-silver ink from the antenna were partially dissolved during pulping and slightly affected the optical and mechanical properties of the recycled paper.

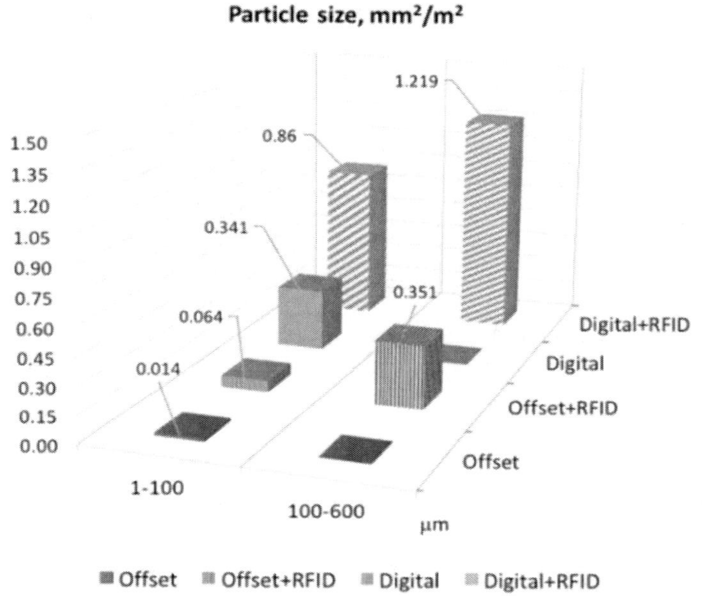

a

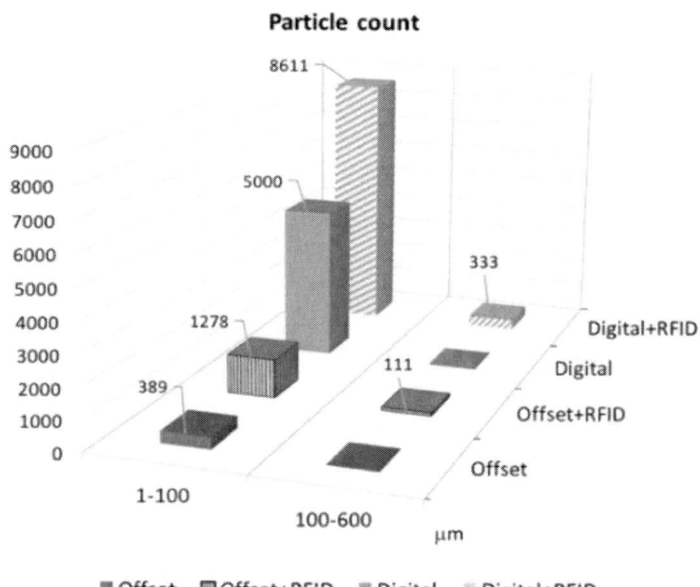

b

Figure 13. Particle size (a) and number of particles (b) of deinked offset and digital prints; equipped with RFID printed antenna and without RFID printed antenna.

In a study performed by the Fibre Box Association the recycling of RFID-tagged corrugated packaging was evaluated (Southwell, 2005). Two types of tags were tested: one containing solid metal (copper foil) and the other a silver ink-printed antenna. Results showed that the laminated solid metal tag was removed intact, whereas the non-laminated tag decomposed into smaller pieces, some of which were too small to be screened out. Tags with printed antennas did not screen out, and most of the silver ended up in the recycling process. Similar results were obtained in another study on the recyclability of corrugated packaging containing laminated solid metal and printed antennas (Wayne, 2009). It was demonstrated that silver ink disintegrates in the process and that the most silver was found in cleaner reject, although some ink particles ended up in the recycled paper.

Vukoje et al. evaluated the deinkability of thermochromic offset inks used for printed temperature indicators (Vukoje et al., 2016). The obtained results indicate that thermochromic prints are poorly deinkable. In flotation deinking, only 12.8% of the total dirt area of thermochromic ink was eliminated. In addition, it was also determined that thermochromic inks can affect process wastewater.

Gregor-Svetec et al. evaluated the recyclability and deinkability of offset and digital printed label papers with integrated printed RFID antennas (Gregor-Svetec et al., 2016). In both cases, label paper was printed with one colour (magenta) and the RFID antenna was screen-printed using a conductive ink based on silver flakes. The assessment of deinkability was performed according to INGEDE Method 11. It was confirmed that the recycling process was not disturbed by the presence of the printed antenna. Good deinkability was determined for both offset and electrographic prints equipped with RFID-printed antennas, with only a small deterioration noted in the optical properties of the recycled paper. On the other hand, in both cases (i.e., offset and electrographic prints), the number of ink particles in deinked recycled paper increased considerably compared to the control sample, a recycled printed label paper without RFID tag, as shown in Figure 13. It was determined that silver particles from the functional ink tend to remain within the deinking process and are not easily extractable from recycled fibres (Figure 14) (Gregor-Svetec,

2016). Gregor-Svetec et al. also evaluated the deinkability of RFID antennas printed with nano-silver functional ink (Gregor-Svetec et al., 2017). Comparison between deinkability of RFID-tagged offset printed label paper using silver conductive ink and nano-silver conductive ink showed better optical properties of recycled paper and higher ink elimination when nano-silver conductive ink was used, although the number of dirt particles attached to recycled fibres after the deinking process was higher.

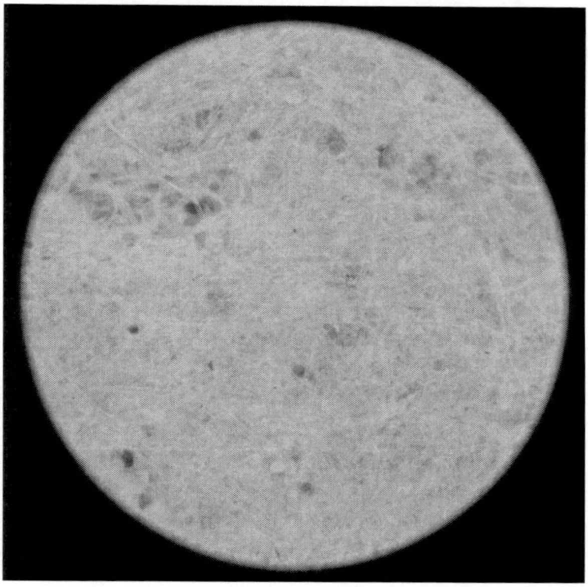

Figure 14. Picture of deinked offset print equipped with RFID printed antenna taken with optical microscope.

5.2. Positive Impact

Despite its environmental drawbacks, RFID could also have a number of beneficial environmental applications. Smart labels are not only contaminants in waste streams; they can also be an opportunity for recycling systems to improve waste management processes (Thoroe et al., 2009).

Item-level tagging can improve the collection process, enabling registration and documentation of collection that can be used for extended producer responsibility and deposit systems. The collection process in deposit systems yields very homogenous waste flows and prevents uncontrolled disposal (Thoroe et al., 2009). Rebates and fees for proper disposal of used items could be initiated, depending on what the consumer puts in the recycling bin. Tags on hazardous products, such as household chemicals, could identify the contents and inform consumers how and where to dispose of them (Thomas, 2008).

Item-level identification enables sorting based on item identity instead of broad physical characteristics, thus leading to a much finer separation of waste fractions. This would also facilitate filtering out hazardous or toxic waste (Thoroe et al., 2009). Tags on batteries, for example, could make the sorting of battery types for recycling cheaper and more efficient (Thomas, 2008). In addition to a higher quality of waste fractions, a higher recycling rate would also be a result of RFID-based waste sorting (Schindler et al., 2012).

Product-related information regarding what should be done with the product at end-of-life (e.g., sell, donate, repair, dispose) or how it should be handled could also be provided by item-label tagging (Schindler et al., 2012). Tags on electronic equipment could be linked to databases on the Web showing how to recycle and dismantle the product, resulting in increased recovery of metals and reusable components (Thomas, 2008). RFID tags could assist in the collection of recyclables in stores or recycling centres and support the reuse, refurbishment and recycling of products. RFID-based information systems can provide data on product history, enabling the rating of material quality for product recovery and stimulating the reuse or redistribution of products with little or no processing (Thoroe et al., 2009). For durable products, a link between production of an item and waste management would allow the reuse of materials. Tags could provide basic information relating to e.g., the composition of the product, which raw materials were used, and which parts can be recycled (Binder and Domnitcheva, 2003).

Item-level tagging could also help consumers to decide which products they want to purchase. Smart labels could provide the consumer with individualised environmental information related to the product (Thomas, 2008). In this way, consumers could make more informed choices based on the sustainability of the products. Consumers could check how far an item has been transported, its dynamic carbon footprint, energy consumption, net greenhouse gas emissions, and so on (Thomas, 2008).

5.3. Future Outlook

It is evident that the influence of smart labels entering waste management processes is varied. In some cases, smart labels have no effect on existing recycling processes (such as aluminium in tin plate processing); however, in most cases they may influence the process or lower the quality of recycled products to an extent that can be quite disturbing, such as with aluminium in glass recycling (Kräuchi et al., 2005). Accordingly, smart labels should be designed with an eye to increased compatibility with existing recycling loops. The materials for production of smart labels should be chosen not only by performance, but also with consideration given to environmental impact. Recyclability can be improved by decreasing the variety of materials in the product. In addition, adaptations should be made to recycling processes in order to prevent tags and components from passing through the recycling process after cleaning and screening. At present, valuable metals in smart labels are irrecoverably lost, mostly by incineration, though the labels can be in some cases collected separately and the components recovered.

Kanth et al. evaluated the sustainability and emissions in the life cycle stages of printed RFID antennas on polymer substrate (Kanth et al., 2012a). They compared PCB technology and polymer substrate-printed RFID technology in terms of input resource utilisation and emissions to air, fresh water, sea water and industrial soil, as well as hazardous emissions. Their study revealed that printed electronic materials are considerably more environmentally friendly than materials needed for PCB electronics, and that tags with printed antennas have less harmful and hazardous impacts on the environment. Kanth et al. also investigated the sustainability, environmental impacts and toxic effects of the manufacturing process for printed RFID antennas on paper substrate (Kanth et al., 2011). The printing technology they assessed was inkjet printing. Again, in this case, they confirmed that printed antennas have a less harmful and hazardous impact on the environment. The comparison between toxic emissions to air, fresh water, sea water and industrial soil in the manufacturing of polymer and paper-based RFID tags showed that paper-based RFID antennas yield lower organic emissions to the environment and considerably higher inorganic emissions to the air (Kanth et al., 2012b).

Compared to conventional, subtractively manufactured electronics, printable smart labels have some advantages where environmental issues are concerned. One of the main advantages of printable electronics is their efficient use of materials, as printing means that only materials that build the smart label are used, which also results in a significantly smaller amount of waste. Another important advantage of printed electronics is the minimal amount of excess chemicals used (Kunnari et al., 2009). However, depending on the ink, nanoparticles and the presence of volatile organic compounds developed from the solvent could have some influence on workers' health and a negative impact on the deinkability of printed labels (Kunnari et al., 2009). Minimising energy consumption in the manufacturing phase and maximising operating life are two problems that should be addressed in the future, in addition to improvement of the recyclability of smart labels.

Conclusion

Instead of conclusion the implementations of information processing and augmented reality in printed smart labels for packaging is in short overviewed.

The packaging of the future will follow the already established trends of sustainability, branding, authenticity and sensing, as well as pursue more robust implementation of artificial intelligence. Consequently, the common conception of packaging as something we "*use and throw away*" will be changed and enriched with the added values of additional uses, such as the transformation of the used package into another useful object, packaging that is edible after use, or even the possibility of combining a number of used packaging pieces into an entirely new entity (Sherman, 2017).

Printed labels and their implementation in packaging will adapt intelligently to new developments in technology. In short, they will follow the trends of their carrier, i.e., packaging, in terms of such properties as bio-degradability and second and further applications in modular use. Moreover, labels will become even smarter with the establishment of more solid links between the physical world and the digital and virtual worlds. Here, developments in the field of dynamic visual media (dynamic web content, dynamic video media, dynamic brand identity), computer graphics (real-time computer graphics), virtual reality (web and mobile applications) and augmented reality (computer-mediated contextual information in the form of visual, auditory and tactile data) are becoming encouraging for printed electronics (Nikobonyadrad, 2012). The crucial role of connectivity between printed devices and virtual worlds will facilitate a level of immersion of consumers in virtual worlds that will become accessible via augmented reality. The experience of perceiving information through auditory, tactile, and visual sensing will be more accurately monitored and interpreted by printed sensors and transmitters in the future (Haller, 2006).

The research fields that present the most opportunities for new virtual implementations are marketing, food products, interior design, clothing, architecture, video games, cultural heritage, surgery and the automotive industry. New approaches in information processing to the interaction between the real and virtual worlds are yielding new meaning and paradigms of communication and relationships between users and products. The application of the virtual world is like an extension of real-world artefacts that are informationally represented and sensorially and cognitively enriched via the opening of non-physical levels of experience (Mattern, 2001).

Consequently, the challenge for developers is to enhance the capability and quality of printed markers and labels so that the mediation between printed electronics and the digital world can become more profound. Moreover, the challenge for designers is to implement new concepts in user-centred design, not only as design principles that follow technological advances, but also as new communication approaches. As smart labels become more intelligent and expand our ability to present data about products, places and environments, technology will enable a more intimate, personal and focused interpretation of digital data by means of enrichment and greater involvement of virtual media (Flanagan et al., 2016).

Concepts such as interchangeability, use and reuse, gamification and augmentation of packaging are already firmly established in the products of Tetra Pack, Ikea, Adidas, Oasis and McDonald's, which are undoubtedly acting as marketing exemplars for other ambitious companies (Dawood, 2016). Developments in the field of data processing facilitated by smart labels will become more interactively computed and perceptually oriented into virtual data interpretations. Data from reality will be mediated via electronic technology and smart labels into a more engaging user experience of the digital/virtual world by means of interactive storytelling, web-based video games, better implementation of high-quality video production, real-time animation, interactive computer-aided simulation and 3D avatars. Moreover, as Abramowitcz contends, printed electronics open new sensory and communication channels to consumers. As the author points out in the discussion, printed electronics enable more profound engagement of the five human senses and a more immersive sensory experience (Abramowicz, 2015).

However, it is also evident that further implementations of augmented reality in smart labelling and packaging present some new challenges and demand special attention, especially as concerns reliability and robustness in applications, health and safety, and advantages by means of concurrence (Sherman, 2017). In their study, authors Mättig et al. present some requirements of new concepts for the technological aspects and information values of interactions between human and computer (machine). The authors explain that these concepts should be designed with reference to more intuitive approaches and with special care taken regarding user perception of the visualisation of information (Mättig et al., 2016).

Furthermore, the abovementioned sensory experience provided by printed electronics and smart packaging should include, in addition to the already firmly established visual information, new approaches to "visual, auditory, tactile, olfactory and gustatory" perceptual data (Abramowicz, 2015). Here, the potential lies in implementation of acoustic enhancement with sound via creative printed electronics on demand, tactile augmentation via perceptually immersive printings of surfaces that can enrich the user's first interaction with and handling of the package, and last but not least, upgraded sensors based on printed solutions that serve the consumers with an engaging sensation of smell and taste before and during the consumption of the products (Abramowicz, 2015).

However, one should not overlook the concerns raised by researchers regarding the future use of printed elements for mediation between the physical and virtual worlds. These concerns involve speculation that these printed elements (as currently used markers, printed labels) could also be restrictive, due to their need to be printed on physical media, when there is a requirement of complex, multi-layered data visualisation and experience. Here, both the already-used applications and new potentials are evaluated and future developments in advanced machine learning and computer vision algorithms and applications are considered (Szeliski, 2010, Ventura and Höllerer, 2016,).

In addition to the technological (e.g., improvements in transistors, resistors, smart indicators and sensors, integrated circuits, LEDs, photovoltaic cells) and design (interaction with mobile applications, implementation of the concept of the "internet of things") aspects that serve as a constant challenge for developers and researchers, the field also requires an up-to-date redefinition and clarification of conceptual frameworks for smart packaging (Yam, 2005). As was already highlighted in earlier studies when the application of smart packaging was in its infancy, the models of packaging functions should be constantly upgraded. As well as the initial, early concepts, when packaging frameworks encompassed basic functions still in use nowadays such as protection, communication, containment and convenience, and modern concepts which consider the more robust, perceptive and intimate implementation and interpretation of digital data, virtual worlds and augmented reality, at least two new functions should be included: that is, networking and context awareness (as well as experience-based functionalities) (Flanagan et al., 2016, Mattern, 2001).

REFERENCES

Abad, E., Palacio, F., Nuin, M., Zarate, A. G., De Juarros, A., Gomez, J. M., Marco, S. (2009). RFID smart tag for traceability and cold chain monitoring of foods: Demonstration in an intercontinental fresh fish logistic chain. *J Food Eng*, 93(4), 394-399.

Abramowicz, D. (2015). *Creating a Sixth Sense for the consumers, Consumer Packaging* (online). (Accessed July 23, 2017). Available from: cantechonline.com.

Active & Intelligent Packaging - Global Market Outlook (2015-2022), Market research consulting. (2016). (online). (Accessed June 15, 2017). Available from: <http://www.strategymrc.com/report/active-and-intelligent-packaging-market/>.

Active, Smart and Intelligent Packaging Market Analysis, Market Size, Application Analysis, Regional Outlook, Competitive Strategies, and Forecasts, 2015 to 2022, *Grand View Research*. (2014). (online). (Accessed June 15, 2017). Available from: http://www.grandview research.com/industry-analysis/active-smart-intelligent-packaging-market/.

Aliaga, C., Ferreira, B., Hortal, M., Pancorbo, M.A., López, J.M., Navas, F.J. (2011). Influence of RFID tags on recyclability of plastic packaging. *Waste Manag,* 31, 1133-1138.

Aliaga, C., Zhang, H., Dobon, A., Hortal, M., Beneventi, D., (2015). The influence of printed electronics on the recyclability of paper: A case study for smart envelopes in courier and postal services. *Waste Manag,* 38, 41-48.

Amin, E.M., Bhattacharyya, R., Sarma, S., Karmakar, N.C. (2014) Chipless RFID tag for light sensing. In 2014 IEEE Antennas and Propagation Society International Symposium (APSURSI); *IEEE: Memphis,* TN, USA, pp. 1308–1309.

Amin, E.M., Bhuiyan, M.S., Karmakar, N.C., Winther-Jensen, B. (2014). Development of a Low Cost Printable Chipless RFID Humidity Sensor. *IEEE Sens. J.,* No. 14, pp. 140–149.

Angeles, R. (2005). RFID Technologies: Supply-chain Applications and Implementation Issues. *Information Systems Management,* 22(1), 51-6.

Atsushi, O., Kentaro, Y. (2006). A temperature-managed traceability system using RFID tags with embedded temperature sensors. *Nec Technical J,* 1(2), 82-86.

Badia-Melis, R., Ruiz-Garcia, L., Garcia-Hierro, J., Villalba, J.I.R. (2015). Refrigerated Fruit Storage Monitoring Combining Two Different Wireless Sensing Technologies: RFID and WSN. *Sensors,* 15, 4781-4795.

Bao, Z. (2000). Materials and fabrication needs for low-cost organic transistor circuits. *Advanced Materials,* 12(3), 227-230.

Bartels, P.V., Tromp, S.O., Rijgersberg, H., Kreft, F., Improvement of the sustainability in the perishable food supply chain by using communicative packaging design. In: J. Trienekens, J. Top, J. van der Vorst, A. Beulens (Eds.), *Towards Effective Food Chains: Models and Applications,* Wageningen Academic Publishers, 2010, pp. 275-291.

Ben Hai, X., Qing-Yao, L., Liang, Y., Run-Ting, F., Zhao-Hui, L., Jia-Rong, P. (2007). A practical web-based tracking and traceability information system for the pork products supply chain. *New Zeal. J Agr Res,* 50(5), 725-733.

Beriain, A., Rebollo, I., Fernandez, I., Sevillano, J.F., Berenguer, R. (2012). A passive UHF RFID pressure sensor tag with a 7.27 bit and 5.47 pJ capacitive sensor interface. In *Proceedings of the 2012*

IEEE/MTT-S International Microwave Symposium Digest, Montreal, QC, Canada, pp. 1–3.

Bibi, F., Guillaume, C., Gontard, N., Sorli, B. (2017). A review: RFID technology having sensing aptitudes for food industry and their contribution to tracking and monitoring of food products. *Trends Food Sci Technol,* 62, 91-103.

Binder, C., Domnitcheva, S., (2003). When Waste Becomes Intelligent: Assessing the Environmental Impact of Microchip Tagging. In *Environmental Assessment in the Information Society* – Joint SETAC Europe and ISIE Meeting, Brussels, SETAC Europe.

Bock, K. (2005). Polymer electronics systems-polytronics. *Proceedings of the IEEE*, 93(8), 1400-1406.

Briand, D., Molina-Lopez, F., Vásquez Quintero, A., Mattana G., de Rooij Nico F. (2012). Printed Sensors on Smart RFID Labels for Logistics. *New Circuits and Systems Conference (NEWCAS)*, 2012 IEEE 10th International, pp. 449-452.

Briand, D., Oprea, A., Courbat, J., Bârsan, N. (2011). Making environmental sensors on plastic foil. *Mater. Today,* 14(9), 416–423.

Bringans, R. D., Veres, J. (2016). Challenges and Opportunities in Flexible Electronics. *Electron Devices Meeting (IEDM),* 2016 IEEE International, 2 p.

Cain, J.T., Lee, K., Integrating sensors and actuators into RFID tags. In S.B. Miles, S.E. Sarma, J.R. Williams (Eds.) *RFID Teachnology and Applications*. Cambridge University Press, 2008, 61-73.

Capitán-Vallvey, L.F., López-Ruiz, N., Martínez-Olmos, A., Erenas, M.M., Palma, A.J. (2015). Recent developments in computer vision-based analytical chemistry: A tutorial review. *Anal. Chim. Acta,* No. 899, pp. 23–56.

Catarinucci, L., Colella, R., Tarricone, L. (2013). Enhanced UHF RFID Sensor-Tag. IEEE Microw. *Wirel. Compon. Lett.,* No. 23, pp. 49–51.

Chaney, R. (2017). Flexible Hybride Electronics Smart Labels. *OPE Journal* No.18, pp. 6-7.

Chang, J., Zhang, X., Ge, T., & Zhou, J. (2014). Fully printed electronics on flexible substrates: High gain amplifiers and DAC. *Organic Electronics,* 15(3), 701-710.

Chang, W.Y.; Fang, T.H.; Lin, H.J.; Shen, Y.T.; Lin, Y.C. (2009). A large area flexible array sensors using screen printing technology," *J. Display Technol.,* 5(6), 178–183.

Chaves, L. W. F., Decker, C. (2010). A survey on organic smart labels for the internet-of-things. In *Networked Sensing Systems (INSS)*, 2010 Seventh International Conference on, 15-18 June 2010, pp. 161-164.

Cho, N., Song, S.-T., Kim, S., Kim, S., Yoo, H.-J. (2005). A 5.1-uW RFID tag chip integrated with sensors for wireless environmental monitoring. In *Proceedings of the 31st European Solid-State Circuits Conference*, Grenoble, France, 12–16 September 2005, pp. 279–282.

Costa, C., Antonucci, F., Pallottino, F., Aguzzi, J., Sarria, D., Menesatti, P. (2012). A review on agri-food supply chain traceability by means of RFID technology. *Food Bioprocess Tech*, 6(2), 353-366.

Crowley, K., Morrin, A., Hernandez, A., Omalley, E., Whitten, P., Wallace, G. (2008). Fabrication of an ammonia gas sensor using inkjet-printed polyaniline nanoparticles. *Talanta,* 77(2), 710-717.

D'Hont, S. The Cutting Edge of RFID Technology and Applications for Manufacturing and Distribution. *Texas Instrument TIRIS*, 2004, 1-13.

Das, R., *RFID Forecasts, Players and Opportunities* 2017-2027. The complete analysis of the global RFID industry. (online). (Accessed July 4, 2017). Available from: http://www.idtechex.com/research/reports/rfid-forecasts-players-and-opportunities-2017-2027-000546.asp.

Dawood, S. (2016). *Sustainable, Changeable, digital – the packaging trends of the future* (online). (Accessed July 12, 2017). Available from: https://www.designweek.co.uk/issues/22-28-february-2016/sustainable-changeable-digital-the-packaging-trends-of-the-future/.

De Donno, D., Catarinucci, L., Tarricone, L. (2014). A Battery-Assisted Sensor-Enhanced RFID Tag Enabling Heterogeneous Wireless Sensor Networks. *IEEE Sens. J,* No. 14, pp. 1048–1055.

De Jong, A.R., Boumans, H., Slaghek, T., Van Veen, J., Rijk, R., Van Zandvoort, M. (2005). Active and intelligent packaging for food: is it the future? *Food Addit Contam Part A,* 22, 975-979.

De La Fuente Vornbrock, A.; Sung, D.; Kang, H.; Kitsomboonloha, R.; Subramanian, V. Delen, D., Hardgrave, B.C., Sharda, R. (2007). RFID for better supply-chain management through enhanced information visibility. *Prod Oper Manag,* 16 (5), 613–624.

Denneulin, A., Blayo, A., Bras, J., Neuman, C. (2008). PEDOT: PSS coating on specialty papers: Process optimization and effects of surface properties on electrical performances. *Progress in Organic Coatings,* 63(1), 87-91.

Denneulin, A., Bras, J., Blayo, A., & Neuman, C. (2011). Substrate pre-treatment of flexible material for printed electronics with carbon nanotube based ink. *Applied Surface Science*, 257(8), 3645-3651.

Directive 2008/98/EC of the European Parliament and of the Council on waste and repealing certain Directives. (2008) *Off J Eur Union* L312/3, p. 3–30.

Directive 2012/19/EU of the European Parliament and of the Council on waste electrical and electronic equipment (WEEE). (2012) *Off J Eur* Union L197/38, p. 38–71.

Directive 94/62/EC of the European Parliament and of the Council on packaging and packaging waste. (1994) *Off J Eur Union* L365/10, p. 0010 – 0023.

Đokić, M., Radonić, V., Mladenović, V., Pleteršek, A., Kavčič, U., Hladnik, A., Crnojević-Bengin, V., Muck, D. (2014). The influence of lamination and conductive printing inks on smart-card operability. *Materiali in tehnologije,* 48(4), 497-504.

Đokić, M., Radonić, V., Pleteršek, A., Kavčič, U., Crnojević-Bengin, V., Muck, D. (2015). Comparison between the characteristics of screen and flexographic printing for RFID applications. *Informacije MIDEM,* 45(1), 3-11.

Eom, K.H., Kim, M.C., Lee, S., Lee, C.W. (2012). The Vegetable Freshness Monitoring System Using RFID with Oxygen and Carbon Dioxide Sensor. *Int. J. Distrib. Sens. Netw,* Vol. 2012, pp. 1–6.

Eom, S. H., Senthilarasu, S., Uthirakumar, P., Yoon, S. C., Lim, J., Lee, C., ... & Lee, S. H. (2009). Polymer solar cells based on inkjet-printed PEDOT: PSS layer. *Organic Electronics*, 10(3), 536-542.

Escobedo, P., Carvajal, M. A., Capitán-Vallvey, L. F., Fernández-Salmerón, J., Martínez-Olmos A., and Palma, A. J. (2016). Passive UHF RFID Tag for Multispectral Assessment. *Sensors*, No. 16, 14 p.

Fernández, I., Asensio, A., Gutiérrez, I., García, J., Rebollo, I., de No, J. (2012). Study of the Communication Distance of a MEMS Pressure Sensor Integrated in a RFID Passive Tag. *Adv. Electr. Comput. Eng.*, Vol. 12, pp. 15–18.

Fernández-Salmerón, J., Rivadeneyra, A., Martínez-Martí, F., Capitán-Vallvey, L., Palma, A.; Carvajal, M. (2015). Passive UHF RFID Tag with Multiple Sensing Capabilities. *Sensors* 2015, No. 15, pp. 26769–26782.

Flanagan, P.; Papadopoulos, D.; Voss, G. (2016). Intimacy and Extimacy: Ethics, Power and Potential of Wearable Technologies, In *Fundamentals of Wearable computers and Augumented Reality*, 2nd ed., Ed. Woodrow Barfield, CRC Press, Taylor and Francis Gorup, p. 31-59.

Furuta, T., Tonooka, H., Kobayashi, M. (2008). Research on Recycling Cardboard with RF Tags/Labels. *The Eight International Conference on EcoBalance*. Dec.10 - 12, 2008, 619-621.

Gaukler, G.M, Seifert, R.W., Applications of RFID in Supply Chains. In H. Jung, F. F. Chen, B. Jeong, (Eds.), *Trends in Supply Chain Design and Management: Technologies and Methodologies*, Springer-Verlag: London, 2007, 29-48.

Ghaani, M., Cozzolino, M., Castelli, C.A., Farris, G.S. (2016). An overview of the intelligent packaging technologies in the food sector. *Trends Food Sci Tech* 51, 1-11.

Gregor-Svetec, D. Intelligent packaging. In: M.A. Cerqueira, J.M. Lagaron, L.M. Pastrana Castro, A.A. Vicente (Eds.) Nanomaterials in Food Packaging: materials, processing technologies and safety issues. *Elsevier,* 2017 (in print).

Gregor-Svetec, D., Muck, D., Pivar, M., Ravnak, D., Bolanča-Mirković, I. (2016). Recyclability of label papers with integrated UHF RFID printed antenna. In: M. Kleebauer, R. Gericke, (Eds.). *Proceedings of joint conference of PTS and COST Action*, PTS Innovative Packaging Symposium 2016, April 6th - 7th, 2016, Munich: PTS, 127-140.

Gregor-Svetec, D., Pivar, M., Bolanča-Mirković, I., (2017) Deinkability of RFID printed antennas. *Project report*, March 2017, 15 p.

Gregor-Svetec, D., Pivar, M., Muck, D. (2015). Properties of recycled papers important for printable antennas. *Cellulose chemistry and technology,* 49(7/8), 701-708.

Guillory, M., Standhardt, G., *NVC world review on supply chain applications of RFID and sensors in packaging*. 2012. (online). (Accessed July 4, 2017). Available from: http://www.nvc.nl/nl/pasteur-sensor-enabled-rfid/.

Haller, M. (2006). Emerging Technologies of Augmented Reality: Interfaces and Design, PA: *IGI Global*, p. 391.

Han, J. H., Seo, G. H., Park, I. M., Kim, G. N., & Lee, D. S. (2006). Physical and mechanical properties of pea starch edible films containing beeswax emulsions. *Journal of Food Science*, 71(6).

Han, J. H., Seo, G. H., Park, I. M., Kim, G. N., Lee, D. S. *J Food Sci.* 2006, Vol. 71, 290-296.

Hasegawa, T., & Takeya, J. (2009). Organic field-effect transistors using single crystals. *Science and Technology of Advanced Materials*, 10(2), 024314.

Hecker, K., Breitung, S., *Organic and Printed Electronics Applications, Technologies and Suppliers*. 5th Edition. VDMA Verlag GmbH: Frankfurt am Main, 2013.

Higgins, S.G., Boughey, F.L., Hills, R., Steinke, J.H.G., Muir, B.V.O., Campbell, A.J. (2015). Quantitative analysis and optimization of gravure printed metal ink, dielectric, and organic semiconductor films. *ACS Appl Mat Interfaces,* 7(9), 5045–5050.

Hong, S.I., Park, W.S. (2000). Use of color indicators as an active packaging system for evaluating kimchi fermentation. *J Food Eng,* 46(1) 67–72.

Hu, J., Zhang, X., Moga, L. M., Neculita, M., (2013). Modeling and implementation of the vegetable supply chain traceability system. *Food Control,* 30(1), 341-353.

Huang, W.D., Deb, S., Seo, Y.S., Rao, S., Chiao, M., Chiao, J. (2012). A passive radiofrequency pH-sensing tag for wireless food-quality monitoring. *Sensors J IEEE*, 12(3), 487-495.

Humbert, A., Tuerlings, B. J., Hoofman, R. J. O. M., Tan, Z.; Gravesteijn, D., Pertijs, M. A. P. (2013). Alow-power CMOS integrated sensor for CO2 detection in the percentage range. 17th Int. Conf. on *Solid-State Sensors, Actuators and Microsystems*, pp. 1649–1652.

IDTechEx. May 2004. *What are Smart Labels? A basic guide to responsive label technologies.* (Accessed 07.05.2017) Available from: http://www.idtechex.com/research/articles/what_are_smart_labels_000 00030.asp.

Ilie-Zudor, E., Kemeny, Z., Blommestein, F., Monostori, L., van der Meulen, A. (2011). A survey of applications and requirements of unique identification systems and RFID techniques. *Comput Ind*, 62, 227-252.

Indicator. *The free dictionary by Farlex*. (2017). (Accessed 10. 7. 2017) Available from: http://www.thefreedictionary.com/indicator/.

Jedermann, R., Ruiz-Garcia, L., Langa, W. (2009). Spatial temperature profiling by semi-passive RFID loggers for perishable food transportation. *Comput Electron Agr,* 65, 145-154.

Jongwoo, S. (2015). End of Paper Labels: Emerging Smart Labels Toward Internet of Things. Internet of Things (WF-IoT), 2015 *IEEE* 2nd World Forum on. 6p. (online).

Kalansuriya, P., Bhattacharyya, R., Sarma, S., Karmakar, N. (2012). Towards chipless RFID-based sensing for pervasive surface crack detection. In *Proceedings of the 2012 IEEE International Conference on RFID-Technologies and Applications (RFID-TA*), Nice, France, 5–7 November 2012, pp. 46–51.

Kanth, R.K., Liljeberg, P., Tenhunen, H., Chen, Q., Zheng, L., Kumar, H., (2012 b) Comparative Toxic Emission Analysis in Production Process of Polymer and Paper Based RFID Tags. 11th *International*

Conference on Environment and Electrical Engineering 18-25 May 2012, Venice, Italy, 184 – 187.

Kanth, R.K., Liljeberg, P., Tenhunen, H., Wan, Q., Amin, Y., Shao, B., Chen, Q., Zheng, L., Kumar, H., (2011) Evaluating Sustainability, Environmental Assessment and Toxic Emissions during Manufacturing Process of RFID Based Systems. 2011 Ninth *IEEE International Conference on Dependable, Autonomic and Secure Computing*, 12-14 Dec. 2011 Sydney, NSW, Australia, 1066-1071.

Kanth, R.K., Wan, Q., Kumar, H., Liljeberg, P., Chen, Q., Zheng, L., Tenhunen, H., (2012a) Evaluating Sustainability, Environment Assessment and Toxic Emissions in Life Cycle Stages of Printed Antenna. *Procedia Eng* 30, 508 – 513.

Kassal, P., Steinberg, I.M., Steinberg, M.D. (2013). Wireless smart tag with potentiometric input for ultra low-power chemical sensing. *Sens. Actuators B Chem.,* No. 184, pp. 254–259.

Kavčič, U., Maček, M., Muck, D. (2015). Ultra-high frequency radio frequency identification tag antennas printed directly onto cardboard used for the manufacture of pharmaceutical packaging. *Journal of imaging science and technology,* 59(5), 50504/1-50504/8.

Kavčič, U., Pavlovič, L., Pivar, M., Đokić, M., Batagelj, B., Muck, D. (2013). Printed electronics on recycled paper and cardboards. *Informacije MIDEM,* 43(1), 50-57.

Kavčič, U., Pivar, M., Đokić, M., Gregor-Svetec, D., Pavlovič, L., Muck, D. (2014). UHF RFID tags with printed antennas on recycled papers and cardboards. *Materiali in tehnologije,* 48(2), 261-267.

Kelly, E.P., Erickson, G.S. (2005). RFID tags: commercial applications v. privacy rights. *Ind Manage Data Syst,* 105(6), 703-713.

Kerry, J.P., O'Grady, M.N., Hogan, S.A., (2006). Past, current and potential utilisation of active and intelligent packaging systems for meat and muscle-based products: A review. *Meat Sci,* 74, 113–130.

Kirchmeyer, S., & Reuter, K. (2005). Scientific importance, properties and growing applications of poly (3, 4-ethylenedioxythiophene). *Journal of Materials Chemistry,* 15(21), 2077-2088.

Kirchmeyer, S., Hecker, K., Breitung, S., Organic and Printed Electronics. Summary – *OE-A Roadmap*. 6th Edition. VDMA Verlag GmbH: Frankfurt am Main, 2015.

Kleper, M.L. Printed electronics and the automatic identification of objects: an investigation of the merging and developing technologies related to the generation beyond print-on-paper. *Graphic Arts Technical Foundation: Pittsburgh,* 2004.

Koutsoumanis, K. (2009). Modelling food spoilage in microbial risk assessment. *Journal of Food Protection,* Vol. 72, No. 2, pp. 425–432.

Kräuchi, P., Wäger, A., Eugster, M., Grossmann, G., Hilty, I. (2005). End-of-life impacts of pervasive computing. *IEEE technology and society magazine* 30, 45-53.

Kress-Rogers, E., Terms in instrumentation and sensors technology, In R. Kress-Rogers, C.J.B. Brimelow (Eds.), Instrumentation and sensors for the food industry, *Woodhead Publishing: Cambridge*, 1998, 673-691.

Kumar, A., Biebuyck, H.A., Whitesides, G.M. (1994). Patterning self-assembled monolayers: applicationsin materials science. Langmuir 10(5):1498–15119. Chou SY, Krauss PR, Renstrom PJ (1995) Imprint of sub-25 nm vias and trenches in polymers. *Appl Phys Lett* 67, 3114–31168.

Kumari, L., Narsaiah, K., Grewal, M. K., Anurag, R. K. (2015). Application of RFID in Agri-food sector-A review. *Trends Food Sci Technol,* 43(2), 144-161.

Kunnari, E., Valkama, J., Keskinen, M., Mansikkamaki, P., (2009). Environmental evaluation of new technology: printed electronics case study, *Journal of Cleaner Production* 17, 791–799.

Kuswandi, B., Wicaksono, Y., Abdullah, J., Heng, L., Ahmad, M. (2011). Smart packaging: sensors for monitoring of food quality and safety. *Sens Instrum Food Qual Saf,* 5(3–4), 137-146.

Lee, C. Y., WuG-Wand, Hsieh, W. J. (2008). Fabrication of micro sensors on a flexible substrate. *Sensors Actuators A: Physical,* 147(1), pp. 173–176.

Lee, S.J., Mijanur Rahman, A.T.M. (2014). Intelligent Packaging for Food Products, In: J.H. Han (Ed.), *Innovations in Food Packaging. Elsevier*: Amsterdam, p. 171-203.

Liquid Crystal Formulations, *LCR Hallcrest*. 2016. (online). (Accessed July 18, 2017). Available from: http://www.hallcrest.com/technology/liquid-crystal/>.

Männl, U., Chuvilin, A., Magunje, B., Jonah, E. O., Härting, M., Britton, D. T. (2013). *Japanese J Appl Phys*. Vol. 52, 05DA11.

Mattern, F. (2001). Linking the Real World to the Virtual World – Atoms, Bits and Smart Objects, lecture Ladenburger Diskurs "Living in a Smart Environment," *ETH Zürich*, 2001. (online). (Accessed July 12, 2017). Available from: https://www.vs.inf.ethz.ch/events/slides/lad2/Mattern2.pdf.

Mättig, B.; Lorimer, I.; Kirks T.; Jost, J. (2016). Analysis of the application of Augmented Reality in the packaging process considering specific requirements on the visualization of information as well as the ergonomic integration of humans into the process. *Logistic Journal, Proceedings*, Vol. 2016. (online). (Accessed July 23, 2017). Available from: http://www.logistics-journal.de/proceedings/2016/fachkollo quium2016/4453.

Meng, Z., Li, Z. (2016). RFID Tag as a Sensor – A Review on the Innovative Designs and Applications. *Meas Sci Rev,* 16(6), 305-315.

Mennecke, B., Townsend, A. (2005). Radio frequency identification tagging as a mechanism of creating a viable producer's brand in the cattle industry. MATRIC Research Paper 05-MRP 8. (online). (Accessed June 15, 2017). Available from: http://www.card.iastate.edu/publications/DBS/PDFFiles/05mrp8.pdf.

Mohebi, E., Marquez, L. (2015). Intelligent packaging in meat industry: An overview of existing solutions. *J Food Sci Technol,* 52(7), 3947-3964.

Molina-Lopez, F., BriandDand de Rooij, N. F. (2012). All additive inkjet printed humidity sensors on plastic substrate. *Sensors Acuators B: Chemical,* Vol. 166-167, pp. 212-222.

Molina-Lopez, F., Vásquez Quintero, A., Mattana, G., Briand Dand de Rooij, N. F. (2013). Large-area compatible fabrication and encapsulation of inkjet-printed humidity sensors on flexible foils with integrated thermal compensation. *J. Micromech. Microeng.,* Vol. 23, 11 p.

Moliton, A., Hiorns, R. C. (2004). Review of electronic and optical properties of semiconducting π-conjugated polymers: applications in optoelectronics. *Polymer International,* 53(10), 1397-1412.

Morrin, A., Ngamna, O., O'Malley, E., Kent, N., Moulton, S. E., Wallace, G. G., Killard, A. J. (2008). The fabrication and characterization of inkjet-printed polyaniline nanoparticle films. *Electrochimica Acta,* 53(16), 5092-5099.

Mraović, M., Muck, D., Pivar, M., Trontelj, J., Pleteršek, A. (2014). Humidity Sensors Printed on Recycled Paper and Cardboard. *Sensors,* No. 14, pp. 13628-13643.

Nair, R., Perret, E., Tedjini, S., Barron, T. (2012). A humidity sensor for passive chipless RFID applications. In *Proceedings of the 2012 IEEE International Conference on RFID-Technologies and Applications (RFID-TA)*, Nice, France, 5–7 November 2012, pp. 29–33.

Nikobonyadrad, S. (2012). Augmented Reality for the Product Packaging, Bachelor Thesis (online). (Accessed July 12, 2017). Available from: http://www.divaportal.se/smash/get/diva2:536726/FULLTEXT01.pdf.

Nilsson, H.-E., Unander, T., Siden, J., Andersson, H., Manuilskiy, A., Hummelgard, M., Gulliksson, M. (2012). System Integration of Electronic Functions in Smart Packaging Applications. *IEEE Trans. Compon. Packag. Manuf. Technol,* No. 2, pp. 1723–1734.

Oprea, A., Courbat, J., Bârsan, N., Briand, D., de Rooij, N.F.; Weimar, U. (2009). Temperature, humidity and gas sensors integrated on plastic foil for low power applications. *Sens. Actuators B Chem.*, No. 140, pp. 227–232.

Papetti, P., Costa, C., Antonucci, F., Figorilli, S., Solaini, S., Menesatti, P. (2012). A RFID web-based infotracing system for the artisanal Italian cheese quality traceability. *Food Control*, 27(1), 234-241.

Park, M., Printed electronics technology and market outlook. *IHS Electronics & Media,* 2013.

Pault, H. (1995). Brain Boxes or Simply Packed? *Food Process* 64(7), 23-24,26.

Pavelkova, A. (2003) Time temperature indicators as devices intelligent packaging. *Acta Univ Agric Silvic Mendelianae Brun* 56, 245-252.

Peng, B.; Lin, J., Chan, P. K. L. (2013). Flexible transistor active matrix array with all screen-printed electrodes," *Proc. SPIE,* vol. 8831, p. 883116.

Pereira de Abreu, D.A., Cruz, J.M., Paseiro Losada, P. (2011). Active and intelligent packaging for the food industry. *Food Rev Int,* 28, 146-187.

Perelaer, J., Hendriks, C. E., de Laat, A. W., Schubert, U. S. (2009). One-step inkjet printing of conductive silver tracks on polymer substrates. *Nanotechnology*, 20(16), 165303.

Pocas, M.F.F., Delgado, T. F., Oliveira, F.A.R., Smart packaging technologies for fruits and vegetables, In: Smart packaging technologies, John Wiley & Sons Ltd., *West Sussex*, 2008, pp. 151-166.

Preradovic, S., Karmakar, N. (2010). Chipless RFID tag with integrated sensor. In *Proceedings of the 2010 IEEE Sensors*, Kona, HI, USA, 1–4 November 2010, pp. 1277–1281.

Pursula, P., Marttila, I., Nummila, K., Seppa, H. (2013). High Frequency and Ultrahigh Frequency Radio Frequency Identification Passive Sensor Transponders for Humidity and Temperature Measurement within Building Structures. *IEEE Trans. Instrum. Meas.,* No. 62, pp. 2559–2566.

Quintero Vásquez, A., Molina-Lopez, F., Smits E. C. P., Danesh, E., Brand, J., Persaud, K., Oprea, A., Barsan, N., Weimar, U., Rooij, N. F., Briand, D. (2016). Smart RFID label with a printed multisensor platform for environmental monitoring. *Flexible and Printed Electronics,* No. 1, 12 p.

Rawat, P., Singh, K.D., Chaouchi, H., Bonnin, J.M. (2014). Wireless sensor networks: a survey on recent developments and potential synergies. *J Supercomput,* 68(1), 1-48.

Realini, C. E., Marcos, B. (2014). Active and intelligent packaging systems for a modern society. *Meat Sci,* 98(3), 404-419.

RFID Tags. 11th *International Conference on Environment and Electrical Engineering* 18-25 May 2012, Venice, Italy, 184 – 187.

Rivadeneyra, A., Fernández-Salmerón, J., Agudo-Acemel, M., López-Villanueva, J. A., Capitan-Vallvey, L. F., Palma, A. J. (2016). Printed electrodes structures as capacitive humidity sensors: A comparison. *Sensors and Actuators A: Physical*, No. 244, pp. 56–65.

Robertson, G.L., *Food packaging: principles and practice*, Marcel Dekker: New York, 1993.

Rodrigues, E.T., Han, J.H., Intelligent packaging. In: D. R. Heldman (Ed.), *Encyclopedia of agricultural, food, and biological engineering*, Marcel Dekker Inc.: New York 2003, p. 437.

Ruhanen, A., Hanhikorpi, M., Bertuccelli, F., Colonna, A., Malik, W., Ranasinghe, D., López, T.S., Yan, N., Tavilampi, M. *Sensor-enabled RFID tag handbook,* 2008, 1-47.

Salmerón, J.F., Rivadeneyra, A., Agudo-Acemel, M., Capitán-Vallvey, L.F., Banqueri, J., Carvajal, M.A., Palma, A.J. (2014). Printed single-chip UHF passive radio frequency identification tags with sensing capability. *Sens. Actuators A Phys*, No. 220, pp. 281–289.

Sangkil, K., Cook, B., Le, T., Cooper, J., Lee, H., Lakafosis, V., Vyas, R., Moro, R., Bozzi, M., Apostolos, G., Collado, A., Manos M. T. (2013). Inkjet-printed antennas, sensors and circuits on paper substrate. IET *Microwaves, Antennas & Propagation, 7(10)*, pp. 858–868.

Sanha Kim, Hossein Sojoudi, Hangbo Zhao, Dhanushkodi Mariappan, Gareth H. McKinley, Karen K. Gleason, A. John Hart. (2016). Ultrathin high-resolution flexographic printing using nanoporous stamps. *Science Advances* 07 Dec 2016: e1601660.

Schindler, R., Schmalbein, N., Steltenkamp, V., Cave, J., Wens, B., Anhalt, A. (2012). Smart trash Study on RFID tags and the recycling industry. RAND Corporation, Brussels, *European Commission*, 277 p.

Schnideritsch, H., Luidold, S., Angerer, T. Fasching, M. J., Antrekowitsch, H. (2012). Problematik der Aufbereitung von RFIDs. in: Thomé-

Kozmiensky, K.J., Goldmann, D. (Eds) *Recycling und Rohstoffe.* Band 5. Neuruppin: TK-Verlag. pp. 625–636.

Sen, L., Hyun, K.H., Kim, J.W., Shin, J.W., Eom, K.H., (2013). The design of smart RFID system with gas sensor for meat freshness monitoring. *Adv Sci Technol Lett* 41, 17-20.

Serenius E. Printed *Electronics and Gravure*. (online). (Accessed August 13, 2017). Available from: http://www.tappi.org/content/events/11papercon/documents/696.632%20A.pdf.

Sherman, L. (2017). *Smart Packaging, SCION*, Lecture 2017, (online). (Accessed August 2, 2017). Available from: http://www.riddet.ac.nz/sites/default/files/content/8%20Smart%20Packaging%20-%20Lou%20Sherman%20.pdf.

Shrestha, S., Balachandran, M., Agarwal, M., Phoha, V.V., Varahramyan, K. (2009). A Chipless RFID Sensor System for Cyber Centric Monitoring Applications. *IEEE Trans. Microw. Theory Tech.,* No. 57, pp. 1303–1309.

Sigma-Aldrich, catalogue. (online). (Accessed August 13, 2017). Available from: (http://www.sigmaaldrich.com/catalog/product/aldrich/476706?lang=en&region=SI&cm_sp=Insite-_-prodRecCold_xorders-_-prodRecCold2-1).

Sigma-Aldrich, catalogue. (online). (Accessed August 13, 2017). Available from: (http://www.sigmaaldrich.com/catalog/product/aldrich/483095?lang=en&region=SI&gclid=EAIaIQobChMI8t3e2vW91QIVSY0bCh14WwPEEAAYAiAAEgITXfD_BwE).

Sirringhaus, H., Kawase, T., Friend, R. H., Shimoda, T., Inbasekaran, M., Wu, W., Woo, E. P. (2000). High-resolution inkjet printing of all-polymer transistor circuits. *Science,* 290(5499), 2123-2126.

Smart Labels Market by Technology (EAS, RFID, Sensing, ESL, NFC tags), Components (Batteries, Transceivers, Microprocessors, Others), Application (Retail Inventory, Perishable Goods, Electronic & IT asset, Others), End-use Industry - Forecast to 2021. Markets and markets.com April 2016. (online). (Accessed June 15, 2017). Available from: http://www.marketsandmarkets.com/Market-Reports/smart-label- market-192651809.html.

Smithers Pira: *Thermochromic inks and reducing household food waste*. 2013. (online). (Accessed July 18, 2017). Available from:<www.wrap.org.uk/food/>.

Smits, E., Schram, J., Nagelkerke, M., Kusters, R., Heck, G., Van Acht, V., Van Koetse, M., Brand, J., Van Den Gelinck, G., Schoo, H. (2012). Development of printed RFID sensor tags for smart food packaging. In *Proceedings of the 14th International Meeting on Chemical Sensors, Nuremberg,* Germany, pp. 403–406.

Smolander, M. *Freshness indicators for food packaging*. John Wiley & Sons Ltd.: West Sussex, 2008, pp. 111–127.

Southwell, J. (2005). *Recycling of RFID-Tagged Corrugated Packaging*. (online). (Accessed July 4, 2017). Available from: http://www.fibrebox.org/upload/2009/Tagged%20Corrugated%20and%20Recycling%20Southwell%20DRAFT%20Presentation.pdf.

Steinberg, M.D., Kassal, P., Tkalˇcec, B., Murković Steinberg, I. (2014). Miniaturised wireless smart tag for optical chemical analysis applications. *Talanta* 2014, No. 118, pp. 375–381.

Street, R.A., Ng, T.N., Schwartz, D.E., Whiting, G.L., Lu, J.P., Bringans, R.D., Veres, J. (2015). From Printed Transistors to Printed Smart Systems. *Proceedings of the IEEE,* 103(4), 607-618.

Sutija, D. (2013). Printed sensor tag and smart objects. *Proc. Energy Harvesting and Storage Europe* (17–18 April Berlin, Germany), pp. 520–530.

Szeliski, R. (2010). *Computer Vision: Algorithms and Applications, Springer*, (online). (Accessed July 22, 2017). Available from: http://szeliski.org/Book/.

Tanaka, H., Matsuki, Y., Shimoda, T., Iwasawa, H., Aoki, T., Yudasaka, I., Furusawa, M. (2006). Solution-processed silicon films and transistors using novel liquid silicon materials. *Digest Tech Papers AM-FPD*, 6, 27-30.

Taoukis, P.S., Labuza, T.P., Time-temperature indicators (TTIs). In: R. Ahvenainen (Ed.), *Novel food packaging techniques*, Woodhead Publishing Limited, Cambridge, 2003.

Thomas, V. M. (2008). *Environmental Implications of RFID 2008 IEEE International Symposium on Electronics and the Environment*, May 19–21, 2008, San Francisco, CA, 1-5.

Thoroe, L., Melski, A., Schumann, M. (2009). Item-Level RFID: Curse or Blessing for Recycling and Waste Management? *Proceedings of the 42nd Hawaii International Conference on System Sciences*, 5-8 Jan. 2009, Big Island, HI, USA, 1-10.

Unander, T., Nilsson, H.E. (2009). Characterization of Printed Moisture Sensors in Packaging Surveillance Applications. *IEEE Sens. J,* No. 9, pp. 922–928.

Vanderroost, M., Ragaert, P., Devlieghere, F., & De Meulenaer, B. (2014). Intelligent food packaging: The next generation. *Trends in Food Science & Technology,* 39(1), 47-62.

Vardeny, Z. V., Heeger, A. J., Dodabalapur, A. (2005). Fundamental research needs in organic electronic materials. *Synthetic Metals*, 148(1), 1-3.

Vásquez Quintero, A et al. "Smart RFID label with a printed multisensor platform for environmental monitoring." *Flexible Print. Electron.* 1 (2016): 025003

Vásquez Quintero, A., Van Remoortere, B., Smits, E. C. P., Van den Brand, J., Briand, D. (2013). Foil-to-foil lamination and electrical interconnection of printed components on flexible substrates *Microelectron. Eng.,* No. 110, pp. 52–58.

Vena, A., Sydänheimo, L., Tentzeris, M.M., Ukkonen, L. (2015). A Fully Inkjet-Printed Wireless and Chipless Sensor for CO2 and Temperature Detection. *IEEE Sens. J.,* No. 15, pp. 89–99.

Ventura, J.; Höllerer, T. (2016). Urban Visual Modeling and Tracking, In *Fundamentals of Wearable computers and Augumented Reality*, 2nd ed. Woodrow Barfield, CRC Press, Taylor and Francis Gorup, p. 174-192.

Voigt, M.M., et al. (2010). Polymer field-effect transistors fabricated by the sequential gravure printing of polythiophene, two insulator layers, and a metal ink gate. *Adv Funct Mater,* 20(2), 239–246.

Vrabič Brodnjak, U. (2017). Experimental investigation of novel curdlan/chitosan coatings on packaging paper. *Progress in Organic Coatings,* 112, 86-92.

Vukoje, M., Jamnicki, S., Rožić, M. (2016). Deinkability of thermochromic offset inks. *Nord Pulp Pap Res J*, 31(4), 692-699.

Wayne K.H., (2009). *The Impact of RFID Tags on the Recycling of Corrugated Containers.* (online). (Accessed July 4, 2017). Available from: http://www.fibrebox.org/upload/2009/PIRArfid.pdf.

Weigelt, K., Hambsch, M., Karacs, G., Zillger, T., & Hubler, A. C. (2010). Labeling the world: tagging mass products with printing processes. *IEEE Pervasive Computing,* 9(2), 59-63.

Whitmarsh, J. (2005). Flexible electronics: silicon meets paper and beyond. *Microelectronics international,* 22(3), 16-19.

Willmann, J., Stocker, D., Dörsam, E. (2014). Characteristics and evaluation criteria of substrate-based manufacturing. Is roll-to-roll the best solution for printed electronics? *Organic Electronics*, 15(7), 1631-1640.

Willmann, J., Stocker, D., Dörsam, E. *Org Electron.* 2014, Vol. 15, 1631-1640. The flexible hybride electronics paradigm (online). (Accessed May 06, 2017). Available from: http://www.uniqarta.com/uploads/3/4/6/4/34643678/the_flexible_hybrid_electronics_paradigm.pdf.

Wojkiewicz, J. L., Bliznyuk, V. N., Carquigny, S., Elkamchi, N., Redon, N., Lasri, T.; Pud, A. A., Reynaud, S. (2011). Nanostructured polyaniline-based composites for ppb range ammonia sensing. *Sensors Actuators B: Chemical,* 160(1), 1394-1403.

Xiao, S., Che, L., Li Xand Wang, Y. (2007). A cost-effective flexible MEMS technique for temperature sensing. *Microelectronics J.,* Vol. 38, pp. 360–364.

Xiao, Y., Yu, S., Wu, K., Ni, Q., Janecek, C., Nordstad, J. (2007). Radio frequency identification: Technologies, applications, and research issues. *Wirel Commun Mob Comp,* 7(4), 457-472.

Yam, K.L., Takhistov, P.T., Miltz, J. (2005). Intelligent packaging: Concepts and applications. *J Food Sci*, 70, 1-10.

Yin ZhouPing; Huang YongAn; Bu NingBin; Wang XiaoMei; Xiong YouLun. (2010). Inkjet printing for flexible electronics: Materials, processes and Equipments, *Chinese Science Bulletin*, October 2010, 55(30), 3383–3407.

Zhang, C. (2014). Printed Electronics: *Manufacturing Technologies and Applications*, (online). (Accessed July 22, 2017). Available from: http://thor.inemi.org/webdownload/2014/Printed_Elec_Zhang_051214.pdf.

Zhu, X., Mukhopadhyay, S.K., Kurata, H. (2012). A review of RFID technology and its managerial applications in different industries. *J Eng Technol Manage*, 29, 152-167.

BIOGRAPHICAL SKETCHES

Urška Vrabič Brodnjak, PhD

Affiliation: University of Ljubljana, Faculty of Natural Sciences and Engineering, Department of Textiles, Graphic Arts and Design

Education: PhD

Business Address: Snežniška 5, 1000 Ljubljana, Slovenia

Research and Professional Experience: Dr. Vrabič Brodnjak teaches as Faculty two subjects: *Graphic Finishing* and *Finishing Graphic Products*. She also has practicum at Packaging and Printing subjects on Bachelor's degree. During 2006 and 2013 she was Head of the Institute for Textile, Graphic Technology and Design in University of Ljubljana. In addition to teaching, Dr. Vrabič Brodnjak is also a member of the Slovenian Institute for Standardisation – Technical Committee for Packaging, member of European Polysaccharide Network of Excellence and also a MC member of COST actions: CA 15126-ENBA, CA 15118-FoodMc and CA15125-Denorms. She recently collaborated on EU project – EcoPaperLoop –

Enhancing the Quality of Paper for Recycling. Dr. Vrabič Brodnjak was and still is involved in many projects, written several scientific papers, participated on many conferences and workshops during her academic profession.

Her research fields are:

- Structural properties and applications of natural, bio-polymers and fibers,
- Natural, bio-polymers, used as an adhesives in the packaging industry,
- Treatments of coating solutions and surface materials, using different technologies (plasma, enzymes, ultrasound),
- Research on the printing properties of coated papers.

Professional Appointments:

Honors: As student she received in 2003, for her diploma thesis, Faculty reward for the best student work/research named Prešernova reward.

Publications from the Last 3 Years:

1. Vrabič Brodnjak, Urška. Experimental investigation of novel curdlan/chitosan coatings on packaging paper. *Progress in organic coatings*, ISSN 0300-9440. (Print ed.), Nov. 2017, vol. 112, str. 86-92, ilustr. http://www.sciencedirect.com/science/article/pii/S030094401730156X. (COBISS.SI-ID 3393904).
2. Vrabič Brodnjak, Urška. Influence of ultrasonic treatment on properties of bio-based coated paper. *Progress in organic coatings*, ISSN 0300-9440. (Print ed.), Feb. 2017, vol. 103, str. 93-100, ilustr. http://dx.doi.org/10.1016/j.porgcoat.2016.10.023, doi: 10.1016/j.porgcoat.2016.10.023. (COBISS.SI-ID 3338608).
3. Vrabič Brodnjak, Urška, Gregor-Svetec, Diana, Klančnik, Maja. Influence of enzymatic treatment on the structural, sorption and dyeing properties of viscose and chitosan/cellulose fibers. *Textile research*

journal, ISSN 0040-5175, 2016, vol. 86, no. 9, str. 990-1005, doi: 10.1177/0040517514559577. (COBISS.SI-ID 3186800).

4. Vrabič Brodnjak, Urška. Microorganism based biopolymer materials for packaging applications: a review. *Journal of composites and biodegradable polymers*, ISSN 2311-8717, Dec., June 2016, vol. 4, no. 1, str. 32-40, ilustr. http://savvysciencepublisher.com/ downloads/ jcbpv4n1a5/. (COBISS.SI-ID 3297648).
5. Vrabič Brodnjak, Urška. UV-drogen van biobased coatings: oppervlakteafwerking voor papier en karton. V: *Papier-gerelateerd onderzoek in Nederland en daarbuiten*, (Innovatieboekjes, 7). Arnhem: Kenniscentrum Papier & Karton (KCPK). Aug. 2016, str. 15, ilustr. (COBISS.SI-ID 3338352).
6. Vrabič Brodnjak, Urška, Muck, Deja. Printing quality of chitosan and rice starch coated packaging paper. V: *Book of abstracts*, 9th National Conference on Chemistry, Sofia, Bulgaria, 29 September - 1 October 2016. Sofia: Faculty of Chemistry and Pharmacy, St. Kliment Ohridski University of Sofia: Union of Chemists in Bulgaria. 2016, str. 87. (COBISS.SI-ID 3298672).
7. Vrabič Brodnjak, Urška. The effect of ultrasound on soy protein and chitosan coating solution for packaging paper. V: *Book of abstracts*, 9th National Conference on Chemistry, Sofia, Bulgaria, 29 September - 1 October 2016. Sofia: Faculty of Chemistry and Pharmacy, St. Kliment Ohridski University of Sofia: Union of Chemists in Bulgaria. 2016, str. 88. (COBISS.SI-ID 3298928).
8. Vrabič Brodnjak, Urška. *Effect of UV-treatment on properties of coated paper with biopolymers: (invited lecture on the European Paper Week 2015, 17th - 19th November 2015, Brussels, Belgium)*. 2015: Confederation of European Paper Industries; Brussels. (COBISS.SI-ID 3200368).
9. Vrabič Brodnjak, Urška. *UV drying for bio-based coatings: (invited lecture on "Industry meets science" / "science meets industry," International paper and board value chain innovation event 2 February 2016, Arnhem, Netherlands)*. 2016: KCPK - Kenniscentrum Papier en Karton; Arnhem. (COBISS.SI-ID 3216496).

Marica Staresinic, PhD

Affiliation: University of Ljubljana, Faculty of Natural Sciences and Engineering, Department of Textiles, Graphic Arts and Design

Education: PhD in information technologies

Business Address: Snežniska 5, 1000 Ljubljana, Slovenia

Research and Professional Experience: printed electronics, image analysis, interactive systems, materials analysis, Photography - *processing and* reproduction of data.

Professional Appointments: Assistant Professor

Publications from the Last 3 Years:

1. Starešinič, Marica, Ahtik, Jure, Učakar, Andrej, Javoršek, Dejana. Evaluation of changes in urban environments using orthophoto maps. V: Cohen, Alexis (ur.). *Urban and built environments: sustainable development, health implications and challenges*, (Urban development and infrastructure). New York: Nova Science Publishers. 2015, str. (41)-53, ilustr. (COBISS.SI-ID 3187056).
2. Javoršek, Dejana, Močnik, Janja, Starešinič, Marica. Analyses of colour appearances on different display devices = Analiza interpretacije barv na različnih prikazovalnih napravah. *Tekstilec: glasilo slovenskih tekstilcev*, ISSN 0351-3386. (Tiskana izd.), 2015, letn. 58, št. 2, str. 100-107, ilustr. (COBISS.SI-ID 3139440).
3. Starešinič, Marica, Javoršek, Andrej, Javoršek, Dejana. Textiles using electronic applications. V: MONDAL, Ibrahim H. (ur.). *Textiles: history, properties and performance and applications*, (Chemistry research and applications). New York: Nova Science Publishers. 2014, str. (447)-463, ilustr. (COBISS.SI-ID 3002992).
4. Ahtik, Jure, Muck, Deja, Starešinič, Marica. A novel database for evaluation of digital images. V: URBAS, Raša (ur.). *Abstracts*, 7th

Symposium of Information and Graphic Arts Technology, Ljubljana, 5-6 June 2014. Ljubljana: Faculty of Natural Sciences and Engineering, Department of Textiles, Chair of Information and Graphic Art Technology. 2014, str. 49. (COBISS.SI-ID 3043184).
5. Starešinič, Marica, Javoršek, Andrej, Javoršek, Dejana. Visokotehnološka oblačila - kaj bomo nosili v prihodnosti? = Hightech dress - what shall we wear in the future?. V: SIMONČIČ, Barbara (ur.), Forte-Tavčer, Petra (ur.). *Trajnostni razvoj v tekstilstvu: zbornik izvlečkov*, 44. simpozij o novostih v tekstilstvu, Ljubljana, 6. junij 2013. Ljubljana: Naravoslovnotehniška fakulteta, Oddelek za tekstilstvo. 2013, str. 48. (COBISS.SI-ID 2880880).

Helena Gabrijelčič Tomc, PhD

Affiliation: University of Ljubljana, Slovenia

Education: PhD

Business Address: Faculty of Natural Sciences and Engineering, Department of Textiles, Graphic Arts and Design, Snežniška 5, 1000 Ljubljana, Slovenia

Research and Professional Experience:
1. A development of *constructional and color-mixing models* based on physical products and computer generated simulations; 2. Generating, processing and *reproduction of data in media* (shape, pattern, color) and three-dimensional computer graphic and visualizations; 3. Research and developing the methodologies of teaching and learning approaches for *web site planning, development and testing and implementation of 2D and 3D computer graphic in creative processes of studying.*

Professional Appointments: Associate Professor

Publications from the Last 3 Years:

Bratuž, Nika, Javoršek, Dejana, Gabrijelčič Tomc, Helena. Defining optimal conditions of colors in 3D space in dependence on gamma values, illumination, and background color. *Journal of imaging science and technology*, ISSN 1062-3701, Jul./Aug. 2015, vol. 59, no. 4, pp. 40503/1-40503/11.

Gabrijelčič Tomc, Helena, Hladnik, Aleš. 1D and 2D shape descriptors applied in fabric drape computer simulation. *Fibres & textiles in Eastern Europe: an international magazine devoted to current problems of the textile industries in Central and Eastern Europe*, ISSN 1230-3666, 2015, vol. 23, no. 6 (114), pp. 92-101.

Arandjus, Mark, Gabrijelčič Tomc, Helena. Rendering 3D graphics as an aid to stylized line drawings in perspective. *JGED: Journal of Graphic Engineering and Design*, ISSN 2217-379X, 2016, vol. 7, no. 2, pp. 5-13.

Iskra, Andrej, Gabrijelčič Tomc, Helena. Eye-tracking analysis of face observing and face recognition. *JGED: Journal of Graphic Engineering and Design*, ISSN 2217-379X, 2016, vol. 7, no. 1, pp. 5-11.

Varga, Luka, Gabrijelčič Tomc, Helena. Subliminal messages in an online environment. *Acta graphica: revija za grafičku tehnologiju, inžinjerstvo i dizajn*, ISSN 0353-4707. (Print ed.), 2016, vol. 27, no. 4, pp. 23-32.

Bratuž, Nika, Gabrijelčič Tomc, Helena, Javoršek, Dejana. Rendering techniques in 3D computer graphics based on changes in the brightness of the object background. V: Cvetković, Dragan (ur.). *Computer simulation*, (Computer and Information Science, Computer Science and Engineering). Rijeka: InTech. 2017, pp. (21)-44.

Kočevar, Tanja Nuša, Gabrijelčič Tomc, Helena. Modelling and visualisation of the optical properties of cloth. V: CVETKOVIĆ, Dragan (ur.). *Computer simulation*, (Computer and Information Science, Computer Science and Engineering). Rijeka: InTech. 2017, pp. (45)-65.

Diana Gregor-Svetec, PhD

Affiliation: University of Ljubljana, Faculty of Natural Sciences and Engineering, Ljubljana, Slovenia

Education: PhD

Business Address: Department of Textiles, Graphic Arts and Design, Snezniska 5, SI-1000 Ljubljana, Slovenia

Research and Professional Experience: characterization of textile and graphic materials, printing substrates and packaging materials, recycling, eco-design of printed products and sustainable packaging

Professional Appointments: Full Professor

Publications from the Last 3 Years:

Scientific articles:

Vrabič Brodnjak, Urška, Gregor-Svetec, Diana, Klančnik, Maja. Influence of enzymatic treatment on the structural, sorption and dyeing properties of viscose and chitosan/cellulose fibers. *Textile research journal*, ISSN 0040-5175, 2016, vol. 86, no. 9, pp. 990-1005, doi: 10.1177/0040517514559577.

Rožić, Mirela, Kulčar, Rahela, Jamnicki, Sonja, Lozo, Branka, Gregor-Svetec, Diana. UV stability of thermochromic ink on paper containing clinoptilolite tuff as a filler. *Cellulose chemistry and technology*, ISSN 0576-9787, 2015, vol. 49, no. 7/8, pp. 693-699.

Gregor-Svetec, Diana, Pivar, Matej, Muck, Deja. Properties of recycled papers important for printable antennas. *Cellulose chemistry and technology*, ISSN 0576-9787, 2015, vol. 49, no. 7/8, pp. 701-708.

Kavčič, Urška, Pivar, Matej, Đokić, Miloje, Gregor-Svetec, Diana, Pavlovič, Leon, Muck, Deja. UHF RFID tags with printed antennas on recycled papers and cardboards. *Materiali in tehnologije*, ISSN 1580-2949, 2014, vol. 48, no. 2, pp. 261-267.

Tišler Korljan, Barbara, Gregor-Svetec, Diana. Properties and printability of compression moulded recycled polyethylene. *Materials & design*, ISSN 0264-1275, Mar. 2014, vol. 55, pp. 583-590.

Zule, Janja, Černič, Marjeta, Gregor-Svetec, Diana. Recyclability determination of paper based packaging products. *Papir: revija Društva inženirjev in tehnikov papirništva*, ISSN 0350-6614, 2014, vol. 42, no. 11, pp. 36-40.

Vodopivec, Jedert, Gregor-Svetec, Diana, König, Silva, Černič, Marjeta. Papir v knjižnem bloku Slave vojvodine Kranjske iz leta 1689 = Paper in the text block of the book The Glory of the Dutchy of Carniola. *Papir: revija Društva inženirjev in tehnikov papirništva*, ISSN 0350-6614, 2014, vol. 42, no. 12, pp. 37-41.

Invited lecture:

Gregor-Svetec, Diana, Zule, Janja, Elegir, Graziano. Recyclability of graphic products and paper based packaging. V: *Wood, pulp & paper 2014: (proceedings of joint conference = zbornik vedeckých konferencií)*, Wood, Pulp and paper 2014 and Polygrafia academica 2014, Bratislava 12.-13. March 2014. Bratislava: Slovenská technická univerzita v Bratislave. 2014, pp. 14-19.

Published scientific conference contribution:

Golja, Viviana, Gregor-Svetec, Diana, Pivar, Matej, Zorič, Andreja, NOVAK, Saša. Detection of silver nanoparticles present in food contact paper with simple screening method. V: *MATBIM 2017, 4th International Meeting on Material/Bioproduct Interaction, Porto, Portugal, 26th-28th April 2017: book of abstracts*. Porto. 2017, str. 86-88.

Gregor-Svetec, Diana, Muck, Deja, Pivar, Matej, Ravnjak, David, Bolanča-Mirković, Ivana. Recyclability of label papers with integrated UHF RFID printed antenna. In: Kleebauer, Markus (ed.), GERICKE, Ralf (ed.). *Proceedings: joint conference of PTS and COST Action*, PTS Innovative Packaging Symposium 2016, April 6th - 7th, 2016, Munich. Munich: PTS. 2016, pp. 127-140.

Heršold, Špela, Gregor-SveteC, Diana. Reusable eco carrier bag from bamboo. In: Simončič, Barbara (ed.), TOMŠIČ, Brigita (ed.), GORJANC, Marija (ed.). *Proceedings*, 16th World Textile Conference AUTEX 2016, 8-10 June 2016, Ljubljana, Slovenia. Ljubljana: Faculty of Natural Sciences and Engineering, 2016, 6 p.

Gregor-Svetec, Diana, Muck, Deja, Pivar, Matej, Đokić, Miloje, KAVČIČ, Urška. Printed electronics on recycled paper, cardboard and foils. In: *Existing technologies and current developments in active and intelligent packaging*, COST Action FP1405 Meeting, Portugal, Aveiro, September 15 -16, 2015. 13 p.

Gregor-Svetec, Diana, Muck, Deja, Pivar, Matej, Golob, Gorazd, Urbas, Raša, Szentgyörgyvölgyi, Rozália, Kavčič, Urška. Novelties in use of recycled paper and paper based packaging. In: Borbély, Ákos (ed.). *Conference proceedings*, International Conference on Design and Light Industry Technologies, November 19-20, 2014, Budapest, Hungary. Budapest: Óbuda University, Faculty of Light Industry and Environmental Engineering. 2014, pp. 23-32.

Gregor-Svetec, Diana, Zule, Janja, Putz, Hans-Joachim, Elegir, Graziano. Recyclability evaluation of paper based packaging. In: URBAS, Raša (ed.). *Proceedings*, 7th Symposium of Information and Graphic Arts Technology, Ljubljana, 5-6 June 2014, 2014, pp. 12-17.

Gregor-Svetec, Diana, Možina, Klemen, Sežun, Mija, Grossmann, Harald. Recycling and waste management in Central Europe countries. In: URBAS, Raša (ed.). *Proceedings*, 7th Symposium of Information and Graphic Arts Technology, Ljubljana, 5-6 June 2014, 2014, pp. 18-23.

Golob, Gorazd, Gregor-Svetec, Diana, Urbas, Raša, Mešl, Mateja, Zule, Janja, Groß, Anja. Assessment of possible improvement of EU legislation on paper recyclability. In: URBAS, Raša (ed.). *Proceedings*, 7th Symposium of Information and Graphic Arts Technology, Ljubljana, 5-6 June 2014, 2014, pp. 37-41.

Gregor-Svetec, Diana, Arko, Manca, Možina, Klemen. Graphic paper from eucalyptus. In: URBAS, Raša (ed.). *Proceedings*, 7th Symposium of Information and Graphic Arts Technology, Ljubljana, 5-6 June 2014, 2014, pp. 218-222.

Deja Muck, PhD

Affiliation: University of Ljubljana, Faculty of Natural Sciences and Engineering

Education: PhD

Business Address: Department of Textiles, Graphic Arts and Design, Snezniska 5, SI-1000 Ljubljana, Slovenia

Research and Professional Experience:

Dr. Deja Muck received her PhD in Graphics in 2002 from University of Ljubljana, Faculty of Natural Sciences and Engineering.

Dr. Muck is a full professor, a lecturer on first- and second-level programs Graphic and Interactive Communication and Graphic and Media Technology and several courses on doctoral programme Textile Engineering, Graphic Communication and Textile Design. From 2006 she also cooperates as a lecturer with University of Zagreb, Faculty of Graphic Arts at doctoral program Graphic engineering and graphic design.

She is authored or coauthored of almost 400 published references, more than 50 (S)SCI articles and other contributions mainly from the fields of paper and print quality assessment, image processing and analysis, printed electronics and 3D-printing with development of 3D-printing materials. She published four books, among the last one was focused on 3D-printing technologies.

Publications from the Last 3 Years

Original scientific articles:

Mraović, Matija, Muck, Deja, Pivar, Matej, Trontelj, Janez, Pleteršek, Anton. Humidity sensors printed on recycled paper and cardboard. *Sensors*, ISSN 1424-8220, Aug. 2014, vol. 14, no. 8, pp. 13628-13643.

Kavčič, Urška, Pivar, Matej, Đokić, Miloje, Gregor-Svetec, Diana, Pavlovič, Leon, Muck, Deja. UHF RFID tags with printed antennas on recycled papers and cardboards = UHF RIFD-značke z natisnjenimi antenami na recikliranem papirju in kartonu. *Materiali in tehnologije,* ISSN 1580-2949. (Tiskana izd.), mar.-apr. 2014, letn. 48, št. 2, pp. 261-267.

Đokić, Miloje, Radonić, Vasa, Mladenovič, Vladan, Pleteršek, Anton, Kavčič, Urška, Hladnik, Aleš, Crnojević-Bengin, Vesna, Muck, Deja. The influence of lamination and conductive printing inks on smart-card operability = Vpliv laminacije in prevodnih tiskarskih barv na delovanje pametnih kartic. *Materiali in tehnologije*, ISSN 1580-2949. (Tiskana izd.), jul.-avg. 2014, letn. 48, št. 4, pp. 497-504.

Majnarić, Igor, Hladnik, Aleš, Muck, Deja, Bolanča-Mirković, Ivana. The influence of ink concentration and layer thickness on yellow colour reproduction in liquid electrophotography toner, *Technical Gazette: Scientific and Professional Journal of Technical Faculties of the University of Osijek* [Utjecaj koncentracije bojila i debljine nanosa žute na kolornu reprodukciju s tekućim elektrofotografskim tonerom. *Tehnički vjesnik: znanstveno-stručni časopis tehničkih fakulteta Sveučilišta u Osijeku*], ISSN 1330-3651, 2015, vol. 22, no. 1, pp. 145-149.

Gregor-Svetec, Diana, Pivar, Matej, Muck, Deja. *Properties of recycled papers important for printable antennas*. Cellulose chemistry and technology, ISSN 0576-9787, 2015, vol. 49, no. 7/8, pp. 701-708.

Kavčič, Urška, Maček, Marijan, Muck, Deja. Ultra-high frequency radio frequency identification tag antennas printed directly onto cardboard used for the manufacture of pharmaceutical packaging. *Journal of imaging science and technology*, ISSN 1062-3701, Sep. 2015, vol. 59, no. 5, pp. 50504/1-50504/8.

Đokić, Miloje, Radonić, Vasa, Pleteršek, Anton, Kavčič, Urška, Crnojević-Bengin, Vesna, Muck, Deja. Comparison between the characteristics of screen and flexographic printing for RFID applications. Information MIDEM*: Journal of Microelectronics, Electronic Sessions in Materials* [Primerjanje lastnosti sito in fleksotiska za aplikacije RFID. Informacije MIDEM: časopis za mikroelektroniko, elektronske sestavne dele in materiale], ISSN 0352-9045, mar. 2015, letn. 45, št. 1, pp. 3-11.

Hvala, Bojana, Simončič, Barbara, Muck, Deja. Vpliv prisotnosti apreture in vode na delovanje pasivnih značk UHF RFID na različnih tkaninah [Influence of finishing and water on functioning of passive UHF RFID tags on different fabrics]. *Tekstilec: glasilo slovenskih tekstil*cev, ISSN 0351-3386. (Tiskana izd.), 2014, vol. 57, no. 2, pp. 153-163.

Pleša, Tanja, Mraović, Matija, Kavčič, Urška, Pivar, Matej, Muck, Deja. Fabrication of printed switches. *Paper technology: the European journal for the fibrous forest products industry*, ISSN 0958-6024, 2016, vol. 57, no. 2, pp. 34-37.

Pivar, Matej, Muck, Deja, Gregor-Svetec, Diana. Primernost recikliranih papirjev za tisk UHF RFID anten [The suitability of recycled papers for printing UHF RFID antennas]. *Anali PAZU*, ISSN 2232-416X, 2016, letn. 5, št. 1/2, pp. 20-25.

Pleša, Tanja, Mraović, Matija, Kavčič, Urška, Pivar, Matej, Muck, Deja. Izdelava tiskanih stikal na papirju [Fabrication of printed switches on paper]. *Papir: revija Društva inženirjev in tehnikov papirništva*, ISSN 0350-6614. (Tiskana izd.), 2016, letn. 44, št. 15, pp. 47-50.

Muck, Deja, Mauko Pranjić, Alenka, Gregor-Svetec, Diana. UV Ink-Jet printability and durability of stone and foil. JGED: *Journal of Graphic Engineering and Design*, ISSN 2217-379X, 2014, vol. 5, no. 1, pp. 7-12.

Igor Karlovits, PhD

Affiliation: Senior Research Associate at the Pulp and Paper Institute

Education: PhD in Graphic engineering and design, University of Novi Sad, Faculty of Technical Sciences, Novi Sad, Serbia

Business Address: Bogišićeva 8, 1000 Ljubljana, Slovenia

Research and Professional Experience: Worked as a associate professor in teaching graphical and printing subjects at the Faculty of technical sciences, University of Novi Sad for more than 10 years. Has been advisers for numerous printing houses, and head of the production at corrugating plant in Slovenia for DS Smith. Currently he is involved in several research projects and working on R&D for fiber based materials and printed electronics.

Professional Appointments: Head of the graphic and packaging department at the Pulp and Paper Institute, Member of the Slovenian Standardization Committee for graphic arts.

Publications from the Last 3 Years:

Jurič, Ivana, Karlovits, Igor, Novaković, Dragoljub, Tomić, Ivana. Comparative study of different methods for the assessment of print mottle. *Color research and application*, ISSN 0361-2317. (Print ed.), 2016, vol. 41, iss. 5,pp. 493-499.

Kašiković, Nemanja, Novaković, Dragoljub, Karlovits, Igor, Vladić, Gojko, Milić, Neda. Colourfastness of multilayer printed textile materials to artificial light exposure. *Acta polytechnica Hungarica*, ISSN 1785-8860, 2015, vol. 12, no. 1, pp. 161-173.

Zorić, Vladimir, Karlovits, Igor. Colour reproduction on tablet devices. *Acta graphica: revija za grafičku tehnologiju, inžinjerstvo i dizajn*, ISSN 0353-4707. (Print ed.), 2014. 25, No. 1/2, pp. 31-36.

Jurič, Ivana, Ranđelović, Danijela, Karlovits, Igor, Tomić, Ivana. Influence of the surface roughness of coated and uncoated papers on the digital print mottle. *JGED: Journal of Graphic Engineering and Design*, ISSN 2217-379X, 2014, vol 5, no. 1, pp. 17-23.

Tomić, Ivana, Karlovits, Igor, Jurič, Ivana. Practical assessment of veiling glare in camera lens system. *JGED: Journal of Graphic Engineering and Design*, ISSN 2217-379X, 2014, vol 5, no. 2,pp. 23-28.

Pavlović, Živko, Dedijer, Sandra, Draganov, Srdjan, Karlovits, Igor, Jurič, Ivana. Offset printing. V: IZDEBSKA, Joanna (ur.), Thomas, Sabu (ur.). *Printing on polymers: fundamentals and applications*. Amsterdam (etc.): Elsevier. 2016, pp. 217-229.

In: Printed Electronics
Editors: K. Yvon and N. Fabrice
ISBN: 978-1-53612-797-3
© 2017 Nova Science Publishers, Inc.

Chapter 2

Electrically Conductive Textile Materials and Printing Inks for Wearable Technology

Maja Klančnik, PhD, Helena Gabrijelčič Tomc, PhD, Diana Gregor-Svetec, PhD, Marica Starešinič PhD and Tatjana Rijavec, PhD
University of Ljubljana,
Faculty of Natural Sciences and Engineering,
Ljubljana, Slovenia

Abstract

This chapter is an overview of technologies, materials, applications and potentials of electrically conductive materials, electrically conductive fibers, conductive printing inks and printing solutions with conductive inks. The wider aspects of the role of conductive materials and inks in wearable technology is presented with the contributions about the environmental impacts of electronic textile wastes and future potentials and concerns about the interconnectivity of physical and virtual world in wearables. In the first part the readers are introduced with the fundamentals of electrical conductivity and resistance of materials. Here, basic definitions are presented, including definitions and equations for electrical conductivity, specific electrical conductivity, electrical resistance, sheet resistance and bulk resistance. Further, the focus is on the electrically conductive materials for the production of conductive textiles, i.e., metals, carbon and conductive polymers. In the next chapters fibres and electrically conductive yarns are shortly explained, with the analysis of the materials used, mechanical and electrical properties and costs of their production. The section of printing technologies includes the explanation of conductive inks, followed by the systematic overview and presentation of different printing technologies (gravure, flexographic, screen and ink jet). The technologies and their applications are also explained with the recently published researches and developments. The chapter continues with the environmental impact of printing ink and proposals for more environmentally-friendly printing solutions. For the understanding of the environmental problem of electronic textile wastes that the society will encounter is further presented the analysis of the production, market demands and waste policies of e-textiles and wearables. The possible solutions for minor environmental impact of electronic textile wastes are proposed with the detailed explanation of eco-design principles and possible solutions with favorable consequences for sustainable developments of e-textiles. The contribution concludes with the future potentials of the use of wearable electronics in textiles. In this part the e-textiles are presented by means of their fundamental role, i.e., the human second skin that should not overlook the basic needs, desires, experiences and expectations of the humans. With some observations about the human sensory system, cognition and the interpretation of the data acquired, measured and processed by e-textiles, the authors predict a new paradigm of design and user-experience of wearables.

Keywords: e-textiles, smart textiles, electrical conductivity, textile materials, printing ink, environmental impact, e-textile wastes

INTRODUCTION

Wearable technology (WT) consists from electronic elements we can wear. WT are electronic devices, like wearable computers, smart watches, smart glasses, smart jewelry etc. or wearable textile systems, such as smart textiles, smart patches and smart clothes. A smart textile with integrated electronic elements, such as a processor, which can control and/or manage information, is called also an electronic textile (e-textile) [1].

WT should be adjusted to a human body shape and its mobility. It can be used for different purposes, like: tracking person's behavior to measure or monitoring physiological and biomechanical functions of human body, tracking environmental factors like temperature, gases etc. by sensors and actuators available, communication, (iv) harvesting energy with photovoltaic devices, etc. (v) heating human body, fashion and design [2, 3].

WT is linked to the World Wide Web for sharing data between the device and the Web over Wife, Bluetooth, GPS or RFID, or it enables to store data and later transfer to other electronic devices. It can be usefully applied in healthcare, medical, public safety, sports, wellness, consumer fitness field, fashion, military, space exploration, etc. [4, 5].

According to Wearable Technology Intelligence Services "the wearable technology industry will see three-fold growth from over $24 billion today to over $70 billion in 2025 [6].

E-textiles become more and more popular in a scientific area and in companies producing highly specialized clothing. Important component of e-textiles are electrically conductive textile materials that are used as conductors of an electrical current or supply of an electrical field to a device [1]. Further, the *electrically conductive fibers* and *conductive printing inks* for integration of planar electronics into textile materials will be presented hereinafter. Effects of printing inks together with technologies and electronic textile waste on the environment are also addressed. At the end are given future potentials and concerns for virtual and augmented aspects of wearable technology on textiles.

Electrical Conductivity and Resistance of Materials

Basic Definitions

Substances have different chemical structure. Their ability to conduct the electric current depends on the number of moving electrons (ions) and temperature. Electrical properties of a substance are given by *electrical conductivity* (*G*) or *electrical resistance* (*R*).

***Electrical conductivity* (*G*)** [S (siemens)] is an ability of the material to conduct the electric current. It is given by the current (*I*) generated by the unit of voltage (*U*) between two points in the electrical circuit (Eq. 1):

$$G = \frac{I}{U} \tag{1}$$

Specific electrical conductivity (γ) [S/m] is a conductivity (G) of a conductor with a length (L) of one meter and a cross-sectional area (A) of one square meter (Eq. 2):

$$\gamma = G \times \frac{L}{A} \tag{2}$$

***Electrical resistance* (*R*)** [V/A =1 Ω (ohm)] is a ratio between the electrical voltage (*U*) and an electrical current (*I*), in accordance with Ohm's law (Eq. 3):

$$R = \frac{U}{I} \tag{3}$$

Specific electrical resistance (ρ) [$\Omega \cdot mm^2/m$] gives an electrical resistance of a wire (conductor) with a cross-sectional area of one square meter and length of one meter. It is a proportionality coefficient between the electrical resistance of the conductor (rod) (R) and the ratio between the cross-sectional area (A) and the length (L) of the conductor (Eq. 4):

$$\rho = R \times \frac{A}{L} \qquad (4).$$

For fabrics and foils, it is more appropriate to apply a *sheet resistance* (R_s *[Ω/□]*. The sheet resistance for a thin film (Figure 1) can be derived by Eq. 5–8:

$$\rho = R \times \frac{d \times W}{L} \qquad (5)$$

$$R = \frac{\rho}{d} \times \frac{L}{W} \qquad (6)$$

$$R_s = \rho/d \qquad (7)$$

$$R_s = R \times \frac{W}{L} \qquad (8)$$

Figure 1. A thin film scheme: *W* – length, *d* – thickness, *L* – wide.

Electrically Conductive Materials

According to the technical report CEN/TR 16298 [1] "an electrically conductive material has a specific conductivity of $>10^{-2}$ S/m (specific resistance $<10^{8}$ $\Omega \cdot mm^2/m$)". The most commonly used electrically conductive materials are metals, graphite, carbon and conductive polymers.

Among *metals*, silver has the highest specific electrical conductivity (63×10^6 S/m), which is more than five times higher than stainless steel (1.45×10^6 S/m) [7]. Specific electrical resistance of metals suitable for use in e-textiles is given in a Figure 2.

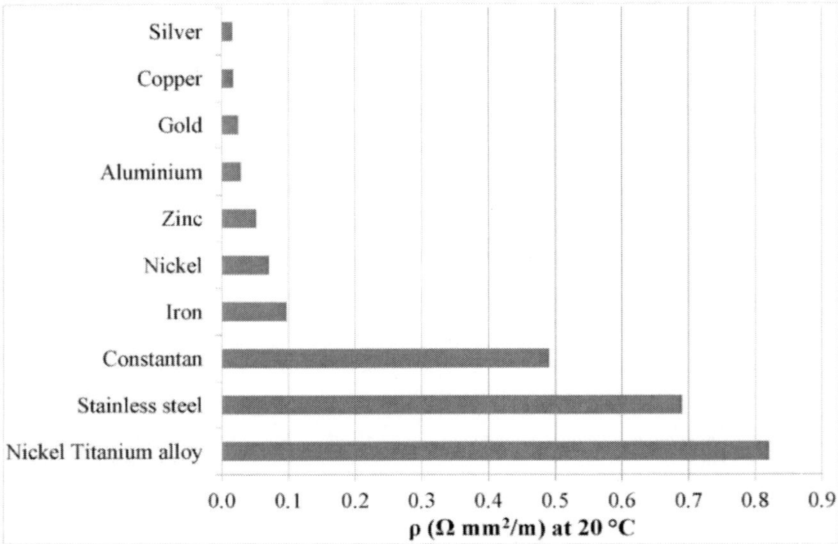

Figure 2. Specific electrical resistance (ρ) of metals at room temperature (adapted from [7, 8]).

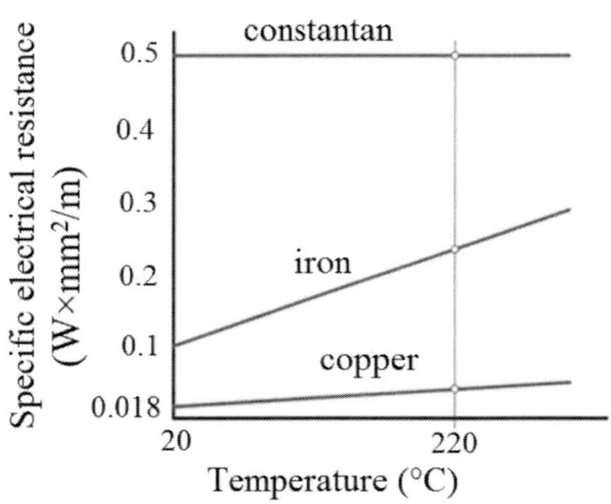

Figure 3. Specific electrical resistance against temperature [9].

Electrical resistance of metals increases with an environment temperature. Increasing of electrical resistance also leads into heating of metals and reduction of electrical conductivity. Constantan, an alloy made from 57% copper, 41% nickel, 1% iron and 1% manganese, is the only metal which electrical resistance does not change with temperature (Figure 3).

Substances from pure *carbon* can exist in amorphous (non-crystalline) form, like coal and carbon black or in different crystal forms, like diamond, graphite, graphene or fullerene. These so-called allotropic carbon modifications arise under different crystallization conditions and have different electrical conductivity (Table 1).

Carbon nanotubes (CNTs) are hollow cylindrical forms with a graphene structure in the sheath, and at the ends they have a structure of fullerene. They are single-layer (SWCNTs) and multi-layer (MWCNTs) [11]. The electrical conductivity of CNTs depend on the chiral angle in the tube structure [12]. CNTs with an "armchair" structure are electrically conductive, better than copper. CNTs with a zigzag structure and chiral structure are semiconductors - they conduct electric current only when additional energy is supplied in the form of light or an electric field, which release some electrons in nanotubes.

Specific electrical conductivity of the CNTs is in range 10–67 x 10^3 S/cm [13] and their specific electrical resistance is in the range 0.1–1 $\Omega \cdot mm^2/m$ [14]. The electrical conductivity of CNTs changes with doping, i.e., binding of other molecules. Doped CNTs are used for making sensors for detection chemical vapours (e.g., carbon monoxide) or biological molecules [15].

Table 1. Allotropic modification of carbon (adapted from [10])

Modification	Electrical conductivity	Applications
Diamond	Poor	Jewellery, blades, etc.
Graphite	Good	Carbon fibres, electrodes, pencils, etc.
Graphene	Very good, comparable to copper	Solar cells (replace silicon), batteries, electrodes, drugs carrier
Fullerene, C60	Semiconductor	Biomedicine

Table 2. Intrinsically conducting polymers (ICPs)

ICPs	Specific electrical conductivity, γ (S/m)	Structure [18, 19, 20]
Polythiofene (PT)	$<10^5$ [21]	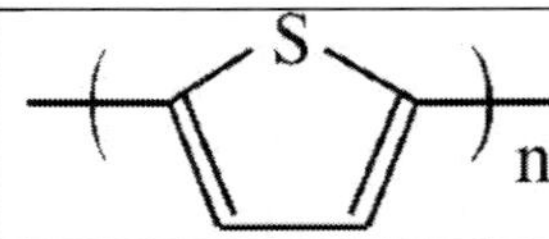
Polypyrrole (PPy)	$200\text{-}10^4$ [22]	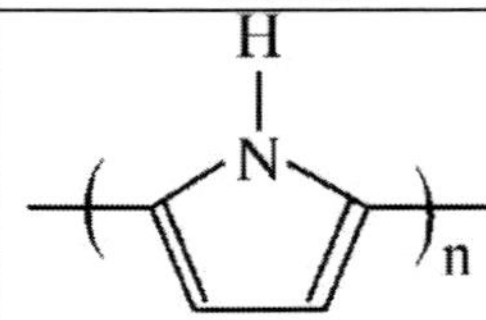
Poly(3,4-ethylenedioxythiophene) PEDOT	3×10^4 [23][a)]	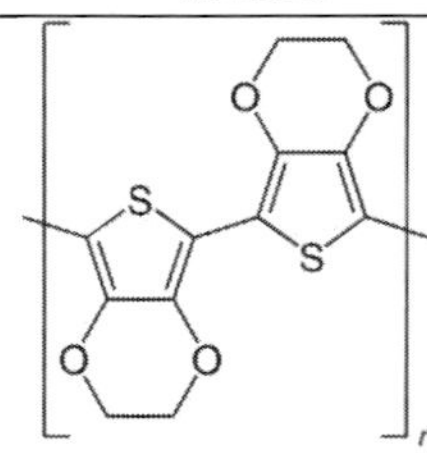
Poly(p-phenylene vinylene) (PPV)	$(1\text{–}1.3) \times 10^3$ [24]	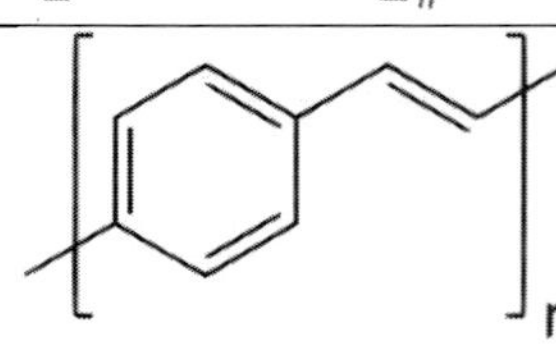
Polyaniline (PANI)	6.28×10^{-9} [25]	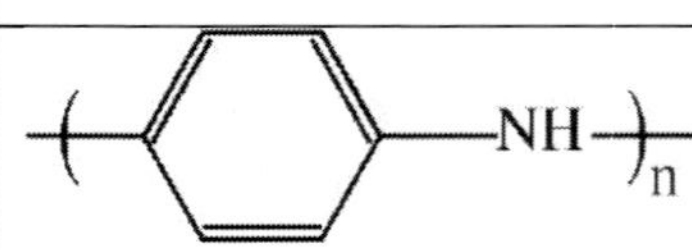

[a)] a doped PEDOT.

Intrinsically conducting polymers (ICPs) are polymers with conjugated double bonds in main chain (Table 2). The conductivity of the intrinsic conjugated polymers without doping is in the range of 10^{-7}–10^{-4} S/m, whereas of common doped conducting polymers is in the range of 0.1–10^5 S/m [16]. Their electrical conductivity is about 40% lower than the electrical conductivity of metals. ICPs, like polyaniline can also reach a metallic conductivity by doping [17].

ELECTRICALLY CONDUCTIVE FIBRES

Highly flexible electrically conductive (electro-active) fibres in yarns or threads can be integrated into textile fabrics by weaving, knitting, embroidery and sewing to make electrically conductive textile materials. Electrically conductive fibres have in electrically conductive textile materials different function. They are used (i) as conductive wires which connect electronic components and a power supply together into an electrical circuit (ii) for transferring data [26], (iii) for generating heat, e.g., for personalize heating and cooling smart clothes [27], for heating thermochromic dyes [28]), (iv) as actuators, e.g., artificial muscles [29], etc. In addition to electrical conductivity, electrically conductive fibres must have good mechanical (tensile, bending, stretching and wrinkling properties) and chemical (resistance to oxidation and to washing [30]) properties.

Table 3. Electrical resistance of textile fibres

Textile fibres	Specific electrical resistance, ρ ($\Omega \cdot mm^2/m$) [31]	Useful as
Acrylic, polypropylene, polyester, polyamide, cellulose acetate and triacetate, aramide, glass fibres, polytetrafluoroethylene	Over 10^{12}	Electrical isolated textile materials
Cotton, wool, regenerated viscose fibres, fibres with special finishing	10^{10}–10^{11}	Antistatic textile materials
Fibres with carbon fillings, mixtures of textile and carbon or steel fibres, textiles coated with conductive polymers	10–10^{9}	Static dissipative textile materials
Metal fibres, carbon fibres	Up to 10	Electroconductive textile materials

Textile fibres differ greatly from one another after specific electrical resistance (Table 3) wherein highly electrical conductive fibres, suitable for electronic textiles, are only metal and carbon fibres. They are intrinsically conductive without adding any conductive substance.

Metal fibers, which are used in electrically conductive textile materials are made from aluminum, copper, nickel, titanium, stainless steel, and from different metal alloys (e.g., nickel titanium, brass). Fine metal fibers are prepared by different methods, like wire drawing, bundle drawing or direct extrusion from the melt [32]. Metal fibers are prepared in continuous form – monofilaments and multifilaments or in discontinuous form - as staple fibers and are processed as conventional yarns [33] *Metal yarns* are made from 100-percent of metal fibers or are prepared by twisting of metal yarn and a non-conductive polymer yarn which improves mechanical properties [33]. Most available are 2-ply or 4-ply yarns. *Metal-wrapped yarns* are combined yarns where a metal filament or metal yarn wraps a strand of non-conductive polymer yarn (Figure 5). *Metal-filled yarns* are another type of combined yarns with a metal filament in the core. The core is covered by non-conductive fibers. Non-conductive polymer yarn/fibres protect the limited elastic metal fibers, which can break easily during manufacturing and use. Non-conductive fibers also provide electrical isolation.

Carbon Fibers and CNTs Yarns

Carbon fibers are highly electrically conductive fibers. Because of high rigidity and brittleness they can be limited used for smart textiles. On the other hand, CNTs can be processed in the same way as conventional fibers, can be drawn and twisted into a yarn [34]. By spinning CNTs together different thickness, strength and conductivity of yarns were achieved. The breakthrough in this area was made by CSIRO Textile and Fiber Technology and the NanoTech Institute of the University of Texas in the year 2006. Ken Atkinson was the first who showed that CNTs could twist like conventional fibers and are capable to self-locking into a yarn. CNTs yarns can be used as mechanical actuators, flexible conductor for textile sensors, power buses for communication, flexible batteries and solar cells, and antiballistic vests [34].

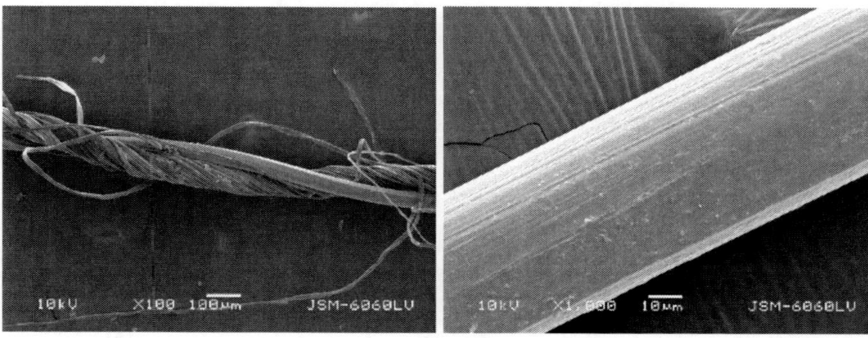

Figure 4. Scanning electron micrographs of a wrapped stainless steel filament with a cotton yarn (left) and its surface view (right).

Coated Textile Fibers by Metals

Textile fibers, in particular manufactured, improve electrical conductivity by coating the surface with metals, galvanic substances or metallic salts. Fibers, mainly metallized with silver, copper, tin and nickel, can be made by vapor deposition, sputtering, reduction of a complex salt of copper, plating without electrodes of noble metals as catalysts [35] and by printing conductive inks.

Coated Textile Fibers by ICPs

ICPs are interesting materials for development electrically conductive fibers due to low density, flexibility, stretchability, excellent adhesion to the textile substrate and non-corrosivity. In spite of good conductive properties ICPs are not used for making fibers because of the poor mechanical properties (brittleness) and high production costs [36]. ICPs don't melt and are insoluble in solvents therefore melt spinning or solvent spinning processes are not suitable for them.

ICPs are used for coating textile fibers and fabrics [37], mainly used for protection against static electricity charge and electromagnetic interference (EMI), but they can also be used for electrically conductive textiles. In comparison to surface resistance of metal coated fibers which is under 1 $\Omega/\square$, the surface resistance of coatings from ICPs, such as PPy and PANI, is in the range from 10 to 10^3 $\Omega/\square$.

Among ICPs, PPy has been mostly used due its high conductivity, low toxicity, commercial availability, and high stability in air [35, 38].

Figure 5. Scanning electron micrographs of a silver coated multifilament synthetic yarn with specific electrical resistance of 4 x 10^3 $\Omega mm^2/m$.

Conductive Polymer Composite Fibers

With adding electroconductive fillers, like carbon black, carbon nanotubes, graphene, metal powders, etc. in the polymer spinning liquid, conductive polymer composite fibers (CPCs) are made.

By increasing the amount of electroconductive fillers, the specific electrical resistance of fibers decreases, but the mechanical properties of the fibers are also deteriorate, which makes the amount of fillers necessary to optimize. A fiber containing electroconduciver fillers is conductive when contacts between the added particles are created. Critical content of filler depends of nature of fillers and their ability to form aggregates and agglomerates which work as current paths [14].

Lim et al., [39] demonstrated a melt spinning process for manufacturing an electrically conductive polypropylene filament yarn with highly conductive silver and CNTs fillers, which have high strength and is suitable for mass production. They found out that added CNTs worked as an electron bridge to form structure for connecting silver particles.

A new strategy to prepare electroconductive fibers was suggested by Han et al., where "first, special thermoplastic polymers are processed into polymer fibers using a conventional melt-spinning process, and then the nonconductive polymer fibers are converted into intrinsically conductive polymer fibers" with iodine doping [36].

Printing with Conductive Inks

Printing technologies are usefully applied to make conductive patterns on the surface of fabrics. They are reasonably priced in comparison with other coating methods such as electroless plating, evaporative deposition and sputtering, where high cost equipment and the use of environmentally undesirable chemicals are required. The printing with conductive inks is also simpler than application of conductive yarns. The highly conductive metal fibers are expansive and their brittle characteristics can even damage spinning machinery over time [40]. However, producing electrically conductive textiles by printing depends on suitable conductive printing inks, their rheological and surface tension characteristics appropriate for selected printing technology and substrate.

Conductive inks must contain a conductive material, which can be metal or non-metallic (as carbon based nanomaterials), a liquid vehicle (aqueous or organic solvents) and various additives (such as rheological and surface tension modifiers, binders, defoamers, etc.) [40, 41, 42].

There are basically three types of metals used in the conductive inks; copper, silver and gold [43]. The gold is very expensive. The copper is less expensive. However, copper oxidizes quickly, making it difficult to work with [44]. The silver is the most electrically and thermally conductive of all metals. Another benefit is that the silver can easily be formulated into inks and the adhesion to substrates is better than others [43]. That is why, the silver based conductive inks are selected to be used for printing.

The *silver inks* are based either on silver precursors or on silver nanoparticles (NPs) [45, 46, 47]. Preparation of silver based inks requires dispersing the silver powder in non-polar organic solvents (usually n-tetradecane) [4]. Tracks printed with metal inks become conductive after thermal sintering, usually at 240-250°C for 30 minutes. Sintering is the process of welding metal particles together at temperature below their melting point and the continuous connectivity between them is achieved [4, 41]. This heat treatment can seriously affect the nature of fibers in textiles, leading to deformation, hardening hand feel and yellowing, especially when thermal sensitive fibers are used. The high temperature also causes oxidation of some metallic nanostructure and results in poor conductivity [48].

There were some approaches to offer alternative sintering techniques, such as laser (photonic), microwave or electrical sintering techniques, instead of heating [48, 49, 50]. These techniques are capable of sintering at relatively low processing temperature but more or less unavoidably involve problems like expensive equipment, complex procedures and not easy large scale production. Some researchers have demonstrated that sintering could be realized at room temperature in the presence of chemical sintering agent [51, 52], such as sodium chloride [53]. In other strategy, hydrogen chloride as sintering agent and polyaniline for improving adhesion were added to silver NPs ink, enabling the low temperature sintering at 60°C for 30 minutes [48]. These researches lead to nowadays developed silver inks, which can be sintered at ambient temperature immediately after contacting the substrate.

The *carbon based nanomaterials,* including carbon nanotubes (CNTs) and graphene, offer a low-cost alternative to the silver inks with excellent environmental stability and desirable conductivity [42, 47]. The carbon inks can also be used in applications where higher resistances are required, providing protection of the silver surface and prevention of silver migration [54].

Graphene is the thinnest and strongest material, also an excellent electrical and heat conductor, made of carbon atoms, arranged in a honeycomb lattice [55]. The unique properties of graphene, such as the high elasticity, mechanical strength, thermal conductivity, very high electrical conductivity and transparency, make them interesting material for stretchable (flexible) electronic applications [4, 56]. It has been found that addition of graphene to printing paste that contains carbon nanotubes significantly enhances the electrical conductivity and sensing properties of the final textile product [56].

A fundamental challenge to integrating carbon nanomaterials with conventional printing technologies was the production of inks suitable for various deposition processes. In recently developed graphene inks, a polymer stabilizer is used to enable stable, high-concentration inks of graphene with tuneable viscosity and solvent composition [47].

There are different methods for printing of conductive inks on the fabrics such as gravure, flexographic, rotary and flat screen printing and ink jet printing [4, 40]. However, printing technologies as inkjet and screen printing (Figure 6) are mostly used for printing electronic patterns on textiles.

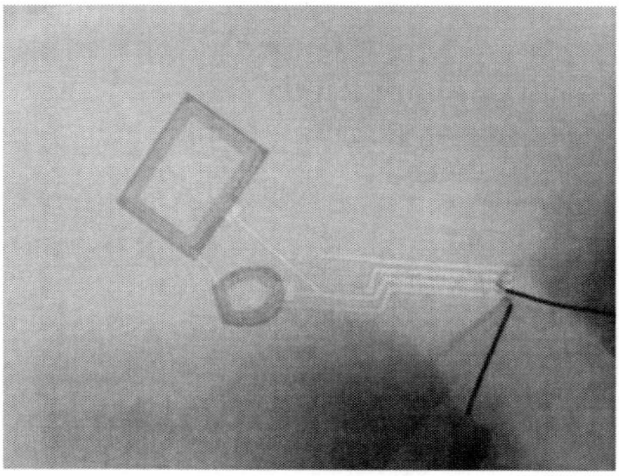

Figure 6. Screen printed electroluminescence pattern on polyester fabric.

Gravure printing is based on the cylindrical printing form (i.e., gravure cylinder) where image elements are engraved into the surface of metal cylinder. Prior the printing, the entire gravure cylinder is inked. Ink is removed from non-image areas by a blade before printing, so that inks remain only in the engraved cells. As the substrate comes in contact with gravure cylinder, the ink transfers from the cells to the substrate by a high printing pressure and creates the print [57]. The making of the gravure cylinder is a long term process including electroplating coatings, engraving the cells and it is also an expensive.

Flexographic printing has the rubber or photopolymer printing plate (or sleeve) mounted on plate cylinder where the image elements are raised above the non-image areas. The raised image elements are coated with ink via an anilox roller. Ink is then transferred from the raised image elements of the printing plate to the substrate by the pressure of the impression cylinder [57]. This method is cheaper than gravure printing. However, the inks costs are significantly higher, because of more required loading of inks to achieve proper shade depth on textile substrates. In addition, this method is frequently slower and it is appropriate for uncomplicated design [4].

Screen printing is a process in which the more viscous ink is forced by a squeegee (blade) through a screen (open mesh) onto the printing substrate. The screen printing plate (Figure 7) consists of the screen fabric mostly from polyester fibers, tensioned on the frame and the stencil containing the printed information. Traditional method of screen printing plate making was coating polyester screen fabric with photosensitive emulsion and exposing it to UV light through positive film. The UV light hardens photoemulsion on non-image areas. The unhardened emulsion is then washed out from image areas of screen mesh. Besides the flat printing screen, the screen can be cylindrical as well, made from nickel sleeve mesh (rotary printing).

Many researchers have used this method to make conductive tracks on the surface of fabric and found out that there are some factors affecting printability such as ink viscosity, mesh count and squeegee hardness [4]. Higher viscosity prevents ink from spreading over the fabric surface and form thicker ink deposit. Lower mesh count deposit more ink while a higher mesh count screen creates an image of higher resolution. A softer squeegee or repeated printing leaves a thicker ink deposit. The thick ink layer is beneficial to electrical performance but disadvantageous to mechanical performance [57].

Ink jet is a non- impact printing technology which does not need the preparation of printing form. It is a clean and environmental-friendly process, which reduces time and costs by increasing productivity of pre-production in printing process [4]. Ink jet printing generates a droplet of ink on the basis of image-dependent signal and sprays this droplet directly onto the substrate [58]. It can be classified as continuous ink jet and drop on demand ink jet that is mostly used.

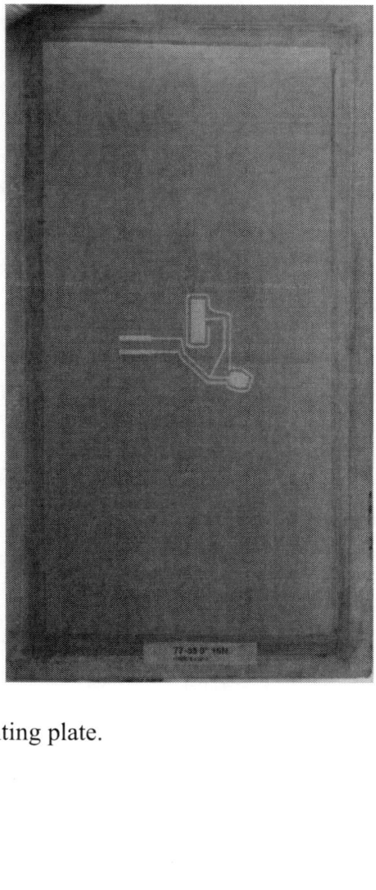

Figure 7. Flat screen printing plate.

The continuous ink jet technology generates a constant stream of small ink droplets, which is charged according to the image and controlled electronically [58]. In drop on demand technology, a droplet is only produced, if it is required by the image. The most important drop on demand technologies are thermal and piezo ink jet. Thermal ink jet generates the drops by heating and localized vaporization of the liquid in a jet chamber. With piezo ink jet, the ink drop is formed and catapulted out of the nozzle by mechanically deforming the jet chamber, an action resulting from an electronic signal and the piezoelectric properties of the chamber wall [58].

Ink jet printing is well suited for low viscosity inks. The dispersion has to be stable and particles must be small enough to fit through the nozzle without clogging. As general rule, the size of the particles in inkjet formulation should be less than 0.1-0.01 of diameter of the pinhead's orifice to avoid its clogging and blockage [42]. The high-viscosity materials, like dispersed particles in inorganic metal inks, cause the nozzle clogging [41]. To overcome disadvantage, a printing solution based on a reactive material which contains silver in form of a dissolved complex and it is therefore particles free was also successfully tested [59].

Conductive printing is optimized when the ink remains mostly on the surface of the fabric and does not penetrate into the internal structure of the substrate. Therefore, flat and smooth surfaces in high-density fabrics woven with fine synthetic filament yarns are more desired for the printing [4].

Most conductive inks are based on silver filler and suffer from brittleness [41]. The problem of their application is to obtain an appropriate thickness of printing ink layer. Another problem is the mechanical durability of printed patterns due to cracking or peeling which occurs in the ink layer. Consequently, a protective nonconductive dielectric layer is necessary to be coated on these surfaces in order to guarantee a long and efficient working life of conductive prints and protection against an electrical short circuit and mechanical damage [4]. Typical conformal coating is made from silicone, polyurethane, epoxy and acrylic resin with different degrees of protection, performance and application. Between them polyurethane coating save conductive prints from losing electrical conductivity after several cycles of mechanical agitation [60].

ENVIRONMENTAL IMPACT

Environmental Impact of Printing Inks and Technologies

Conductive inks based on carbon and graphene are more environmentally friendly because they do not contain metal nanoparticles, which are harmful to our health and environment. Silver is also not recyclable, while graphene and other carbon materials can easily be recycled.

The inks containing water instead of organic solvents are more environmentally friendly. The inks containing organic solvents cause the harmful emissions of volatile organic compounds (VOCs) during the printing and drying the prints, which also lead to formation of harmful tropospheric ozone and smog. The washing off these printing inks from printing equipment requires also organic solvents, which leads to another harmful VOCs emissions and harmful wastes. The water based inks can be washed off with water. However, the polluted wastewater needs to be properly treated before discharging to sewage system.

Comparing environmental impacts of printing technologies shows that inkjet printing is the most environmentally clean printing process. It does not need prepress preparation involving platemaking, which reduces amount of wastes as well as printing time and costs.

When the conventional printing technologies are applied, besides the use of the environmentally less harmful inks, the digital platemaking (computer to plate/cylinder) should be used. In computer to plate (CtP) technology, the phase of producing of photographic film is eliminated and therefore we do not have problems with manipulation with the toxic developing agent, harmful fixation agent, dangerous wastes and polluted wastewater. In the flexography, the digital photopolymer printing plates, which process with water not with organic solvents or even only with heat without any chemicals have to be used. In gravure printing, the gravure copper or zinc cylinders, which are produced by direct laser engraving without chemical etching are better be used. In preparation of printing plate for screen printing, the sensitive photoemulsion (like diazo) without any harmful substances and the computer to screen (CtS) for lower environmental impact must be applied.

Environmental Impact of Electronic Textile Wastes

Steady growth of global textile consumption, with expected growth at a CAGR of 5.4% through 2020, and average lifetime of 3.3 years for clothing items and other textiles, requires textiles and clothing sustainability [61, 62]. Recycling strategies that comply with the circular economy principles are needed, especially because large amounts of clothing and other textiles are discarded every year, nearly 10 million tones in the United States [63]. While approximately 75% of the pre-consumer textile waste is recycled, most of the post-consumer textile waste goes to landfill or is burnt in municipal waste incinerators. In the European countries 5.8 million tones of textiles are discarded every year, with only 25% of these post-consumer textiles being recycled by charities and industrial enterprises [64]. According to EPA, the textile recycling industry in the United States recycles only 15% of post-consumer textile waste each year [65]. Rather than disposing clothes and textiles, reusing and recycling them would be more appropriate, not only because of environmental reasons, but also from economic and social point of view [66].

In the future, the mass production and application of electronic textiles can be expected. Wearable technology has witnessed a growth rate of around 126% during 2014-2015 and is expected to continue this positive trend in the future [67]. North America dominates the global market owing to the rising adoption of e-textile products in the regional healthcare and sports sector. Research by IDTechEx reveals the projected growth for the e-textiles market from a base of $100 million in 2015. The US market for e-textiles, or smart clothing, is projected to reach over US$ 3 billion by

2026 [68]. Most of e-textiles will be used in the sports and fitness sector, military, compression wear, medical, healthcare, and workwear.

Electronic products, also e-textiles usually have rather short service life. The fashion trends are changing; therefore e-textiles can soon become obsolescence. Disposal of e-textiles after their useful life could cause a waste problem, especially if they will be mass produced. It can be expected that large amounts of discarded e-textiles will cause recycling and disposal problem if they are designed without taking into account their end-of-life options [69]. According to some experts, focus on environmental impacts, energy and resources saving, and recycling will be a mega-trend influencing the design of e-textiles [70, 71]. Eco-design focused on the prevention on waste and design for recycling aiming at the optimization of products for repair, refurbishment, and recovery of the materials after usages should not be overlooked at the development process of e-textiles [70, 72].

E-textiles can be divided into e-textiles using classic electronic devices, such as wires, batteries, integrated circuits, LED's, and e-textiles with integrated electronic components, such as fibre based electrodes (sensors and actuators), woven, embroidered, or printed logic circuits, flexible lighting elements, components for power generation and storage (e.g., flexible solar cells, batteries) [70, 73]. From an environmental impact point of view e-textiles have some serious drawbacks, such as high energy consumption, they contain a variety of harmful substances, nonrenewable materials, such as precious metals and rare earth elements, which leave a large environmental footprint even before they are incorporated in products, as well as expected recycling difficulties [74].

In Europe, design of products focused on waste prevention is encouraged with the waste policies. A hierarchy of waste management strategies where waste prevention has priority over reuse, recycling, recovery, and disposal is specified in the Waste Framework Directive [75]. The European RoHS directive stipulates phasing out certain hazardous substances from EEE, such as lead, mercury, cadmium, hexavalent chromium, poly-brominated biphenyls, and polybrominated diphenylether [76]. Design for recycling is encouraged by the European WEEE directive

[77]. At present, WEEE directive doesn't applies to e-textiles yet, however if large quantities of e-textiles will end in waste streams it is expected that e-textiles will be included as products that fall into its regulation. If batteries, electronic devices, electrical contacts are incorporated into textile they must be removed before textile disposal. Waste batteries are regulated by European Directive, and they should be recycled as other electronic and electrical wastes [78].

The biggest fraction of discarded e-textiles is expected to be disposed with municipal solid waste. The small electronic devices integrated in these textiles will be difficult to collect and recycle, because these tiny devices will be scattered in general household waste streams. These small e-waste items could find their way from the household waste into recycling process of non-electronic wastes, where they can act as contaminants, disturbing recycling process, realising even toxic substances [69].

The municipal solid waste can be disposed by incineration or by direct landfilling, depending on the country. The incineration of e-textiles would disperse the voluble metals (such as silver) in the bottom ashes or in the filter dust, which are latter disposed as hazardous waste [69]. The landfilling of e-textiles could cause contamination of soil and groundwater.

E-textiles can be produced by adoption (e.g., pockets) or integration (e.g., embroidered) of electronic devices into textile substrate. In the textile and apparel sector environmental issues were addressed mainly for the manufacturing processes of textiles and clothes, whereas design for recycling and post-consumer waste were not considered as important. For example, at design of e-textiles the trend is to integrate electronic devices seamlessly, which is opposite to the design for recycling for easy disassembly. Another example is the use of metal-coated yarns that is against the design for recycling, which limits the use of plastic materials with surface metallization [70].

Eco-design and design for recycling for e-textiles is difficult task, because of combination of completely different materials and their functions, different use-cycles and end-of-life treatments [70].

Table 4. Eco-design principles that represent challenges for sustainable design of e-textiles (adapted from [70])

Eco-design principle	Possible solution
Eco-design principle referring to material efficiency	
Avoid/reduce different materials in the product.	Avoid different fibers for textile substrates; Use conjugated polymers, organic electronics free of metals and silicon for electronic components.
Reduce weight/thickness of the product.	Use lightweight textiles, thin-layer electronic components, miniaturized ICT-devices.
Use renewable/recycled materials.	Use natural fibers and recyclable materials.
Eco-design principle referring to hazardous substances	
Avoid use of materials that cause adverse environmental impact.	Follow the RoHS directive; Identify and phase out hazardous substances; Use less hazardous alternatives; Replace silicon-based electronics with polymer-based.
Disposal of batteries - easily identifiable/removable.	Design e-textiles in the way that waste batteries are collected and recycled in an ecologically sound way.
Provide information about components requiring special handling.	Inform end-users, collectors, recyclers what kind of hazardous substances are present, how to separate them and how to recover or recycle them.
Eco-design principle referring to end-of-life-treatment	
Recyclability/biodegradability.	Use biodegradable and recyclable fibres for textile substrates; Use recyclable materials for e-components.
Disassembling and separation of components/materials.	Design for easy and safe separation of electronic components from textile substrates, without damaging them substantially; Develop methods and ways to identify electronic components incorporated in textiles.
Avoid/limit different materials, coatings and surface finishes. It could be fallowed only to the point, where functionality of e-textiles is maintained.	Use textiles for housing and packaging of gadgets; Mount electronics on textiles with accessory parts made from textiles.
Stipulate long-term usage, reuse/refurbishment.	Design e-textiles in the way that design is timeless, enabling to adjust design features to fashion trends; Allow upgrade and repair; Use standardized electronic components when applicable; Avoid usage of incompatible electronic components; Provide solution to update wearable computing devices; Make available servicing information to preserve smart functions.

In Table 4 some of eco-design principles related to design for environment and waste prevention that could be applied and represent challenges to e-textiles are shown. It seems, that for e-textiles the waste prevention and solutions which lead to prolonging their useful life are the best option. One of these solutions provides printing technology, which could be applied to print-over faulty organic electronic components [70]. Eco-design principles from the textile and electronics sector must be developed further to avoid cross-contamination during recycling and dispersion of valuable materials within waste streams. Preventive measures for easy waste managing and recycling of e-textiles must be planned at the early development phase of design and integration of electronic and electrical elements into textiles.

Future Potentials and Concerns for Virtual and Augmented Aspects of Wearable Electronics on Textiles

From the technical and technological point of view the developments are directed towards the researches in the field of nanomaterials and nanoelements, improvements of conductivity of yarns and polymers (electrical, magnetic), improvements of the quantity and quality of the memory and phase-changing properties of materials and improvements and specialization of the self-cleaning and antimicrobial properties of the materials [41]. Besides also the technical and perceptive aspects of the interpretation of digital and virtual data that are presented by printed electronic wearables should not be overlooked. Nevertheless, when studying the field one can't avoid the more subtile challenges, which the developers are faced with, when wearable electronic is used, i.e., the known and hidden aspects of the users immersion in the implemented meaning of the data presented by wearables and the introduction of augmented reality [79]. As the human understanding of itself, world and nature is becoming more multi-dimensional, also the media, which the human is interacting with, including wearable devices, should offer a multi-dimensional, therefore multi-layered experience [80]. As the printed wearable electronic on textile and cloths are the enrichment of the human bodies and objects as interiors and provide added value to the wearer and objects, also the output data should be more focused not only on the raw communication of the data as psychophysical state of the body, temperature, etc. but also on the relation between the degree of interconnection of the reality and the virtuality [81]. This relation is going to be changed in the future solution. The predictions of the immersion of the virtual worlds in wearables tend to be usually exaggerated, however, some of the potentials that were science fiction not long ago, have been already realized and launched on the market [82]. By all means, the actual and future solutions should not overlooked the basic need of human beings, that is staying in connection with his/her self, his/her needs, desires and ambitions and the relationship and communication with the soundings (social, environmental). Due to the tendency of new technologies and media of networking, connection and sharing the challenge of keeping appropriate ethics, intimacy and consistency of the data and their interpretation and presentation become crucial. As the electronic textiles, especially when they are printed, what in generally reduce their physical manifestation on (in) textile, cloth and garment, are an optimal wearable bridge for more direct virtual experience and users' immersion into it, the designers and textile technologist should focus on a new paradigm of life style, business and social relations that a modern man is going to established in the next decades [82, 83]. As researches have confirmed, the high number of the users have lost their confidence and interest for the wearable devices after a quite short time of the use. The cause can be probably found in the focus of the developers that with the ambition of the offering to the users an exclusive experience, have not thought about the fact that wearable devices should be in the first place completely comfortable and wearable [84]. Besides the question is also how the human cognition is changing when exposed to continuous interaction with interfaces and electronic devices. Here, the developers should be aware that probably the users get sooner tire off with the data that are presented in electronically and virtually enriched experience in comparison with the experience of self and inner established self-relationship and balance [85]. Nevertheless the data presented by smart electronic devices should help the users to come back to themselves, their feelings, their family, social values and nature [86].

The humans have very intimate relationship with the garments and cloths and the solutions of future wearable electronic in textile should focus on more personal interpretation of human's body properties (»rhythms, movements and biological impulses« [83] and the potentials of interconnectivity between real and virtual world that can enable new aspects of user experience and user engagement [87, 88].

The potentials of printed wearables are the development of the micro-electronics, printed sensors, conductors, antennas [89, 79], the enhancement of internet of things and the more subtile bridging in-between the reality and virtuality with the new semantic of human-computer communication. Here, the virtual information is going to more firmly establish its role in the applications of wearable technology and e-textiles. Besides, the virtual solutions generated by wearable technology are going to mediate the data from real world more intimately and with more focus (human body, five senses, motion, cognitive processes). Consequently, the creation of more engaging user-centered output information is going to be possible [81].

Due to rapid developments of the technology, printing electronics in textile is going to enable more resistible implementations of sensors, actuators, microphones, detectors, smart markers, integrators etc. [41] which are going to transform and alter the real data and interpret in augment data with unique personal, social and environmental imprint [90] According to iQ research [91, 92] the future of wearable tech is step by step built today with the understanding of human behavior and interaction with real (his/her body, environment) and virtual (new media) world.

Wearables are contributing with to man's everyday life with Bio-Tech fusion, Synced Lifestyle, Organic Computing, Human Enhancement, Health Empowerment and Personalized Context. Interactions human to human and human to computer are going to be established with: 1.

connected intimacy that is by authors defined with long distance togetherness, data-streamed care and emotional mirroring, 2. tailored ecosystem (human-computer interaction) that is defined with biotechnology, biometrical attenuation and responsive coaching and 3. coevolved possibilities enabled by augmented sensory perception, on board interfaces, authenticated self and cloud memory [91, 92].

Conclusion

The potentials of printed wearable electronics that can be smoothly implemented in the garment and technically can mimic human skin as a second skin that augments the human experience is a parade horse for modern textiles and clothing. Certainly, the challenges and discussions of how the wearable technology can be, depending on a context of use, on one hand unnoticeably and on the other hand distinctive introduced in fashion, tech textiles and interior and exterior textiles are and are going to be still present, however the focus is pointed on the level and abundance of communicativeness that the printed wearable technology can enable including the immersion of user-centered experience of digital data presented by devices and reinterpretation in the experience of virtual worlds.

References

[1] CEN/TR 16298. (2011). *Textiles and textile products - Smart textiles - Definitions. Categorisation, applications and standardisation needs*, doi: 10.3403/30245031u.

[2] Stoppa, M., & Alessandro, C. (2014) Wearable Electronics and Smart Textiles: A Critical Review. *Sensors*, 14, 11958–11992, doi.org/ 10.3390/s140711957.

[3] Röpert, A. (2013). "Wearable electronics 2.0". *Avantex Symposium* 2013, June 11–13, Frankfurt am Main.

[4] Honarvar, M. G., & Latifi, M. (2016). Overview of Wearable Electronics and Smart Textiles. *J Text I*, 108, 631–652. doi: 10.1080/00405000.2016.1177870.

[5] Koncar, V., editor. (2016). *Smart textiles and their applications*. 1st ed. Amsterdam: Woodhead Publishing.

[6] Wearable technology intelligence services. (2017). Available from: https://technology.ihs.com/Services/511880/wearable-technology-intelligence-service. Accessed: August 17, 2017.

[7] Electrical resistivity and conductivity. Available from: https://en.wikipedia.org/wiki/Electrical_resistivity_and_conductivity. Accessed: August 18, 2017.

[8] Physical properties of nitinol. Memry a SAES Group. Available from: http://memry.com/nitinol-iq/nitinol-fundamentals/physical-properties. Accessed: August 19, 2017.

[9] eOET-1 plus. Available from: http://eoet1.tsckr.si/plus/inc/02_04_01_04/slika_2-4-1-4-2_2.html. Accessed: August 18, 2016.

[10] Kok, L. (2014). Tutorial on allotropes of carbon, metallic bonding, properties of graphene and alloys. Available from: http://www.slideshare.net/wkkok1957/ib-chemistry-on-a. Accessed: August 17, 2017.

[11] Reilly, R.M. (2007). Carbon Nanotubes: Potential Benefits and Risks of Nanotechnology in Nuclear Medicine. Available from: http://jnm.snmjournals.org/content/48/7/1039/F1.expansion.html. Accessed: February 2, 2016.

[12] Boysen, E., & Muir, N.C. (2005). *Nanotechnology for Dummies*. Hoboken: Wiley.

[13] Lekawa-Raus, A., Patmore, J., Kurzepa, L., Bulmer, J., & Koziol, K. (2014). Electrical Properties of Carbon Nanotube Based Fibers and Their Future Use in Electrical Wiring. *Advanced Functional Materials*, 24, 3661–3682. https://doi.org/10.1002/adfm.201303716.

[14] Köhne, N. Electrical conductive viscose fibres and their applications. 53th Man-Made Fibers Congress, Dornbirn 2014. Available from: http://www.kelheim-fibres.com/pdf/2014_Presentation_Dornbirn_

[15] NKoehne_Electrical_Conductive_Viscose_Fibres_and_their_Applica tions.pdf. Accessed: August 18, 2017.

[16] Miao, M. (2011). Electrical conductivity of pure carbon nanotube yarns. *Carbon*, 49, 3755–3761, doi.org/10.1016/j.carbon.2011.05.008

[17] Li, Y. (2015). Conducting polymers. In: *Organic optoelectronic materials*. Edited by Yongfang Li. Series Vol. 91, 25–30. Cham: Springer International Publishing, doi: 10.1007/978-3-319-16862-3.

[18] Posudievsky, O. Y., et al., (2014). Metallic conductivity of mechanochemically doped polyaniline. *Theor Exp Chem*, 50, 197–203, doi: 10.1007/s11237-014-9373-7.

[19] Barron, A. R. (2009). Polymer light emitting diodes. Available from: http://legacy-cte-cnx-dev.cnx.org/content/m25670/1.2/?collection=col10719/latest. Accessed: August 17, 2017.

[20] Shaddack. (2005). Available from: https://commons.wikimedia.org/w/index.php?curid=448109. August 17, 2017.

[21] Yikrazuul. (2008). Available from: https://commons.wikimedia.org/wiki/File:Polyphenylene_vinylene.svg#/media/File:Polyphenylene_v inylene.svg. Accessed: August 17, 2017.

[22] Polythiophene. Available from: https://en.wikipedia.org/wiki/Polythiophene, Accessed: August 17, 2017.

[23] Polypyrole. Available from: https://en.wikipedia.org/wiki/Polypyrrole. Accessed: August 17, 2017.

[24] Le, T.-H., Kim, Y., & Yoon, H. (2017). Electrical and electrochemical properties of conducting polymers. *Polymers*, 9, 1-32, doi: 10.1007/978-94-011-1568-1_1. Available from: file:///C:/Users/tarijavec/Downloads/polymers-09-00150.pdf. Accessed August 17, 2017.

[25] Poly(p-phenylene vinylene). Available from: https://en.wikipedia.org/wiki/Poly(p-phenylene_vinylene). Accessed: August 17, 2017.

[26] Polyaniline. Available from: https://en.wikipedia.org/wiki/Polyaniline. Accessed: August 17, 2017.

[27] Rantanen, J., Hänninkäinen, M. (2005). Data transfer for smart clothing: requirements and potential technologies. In: *Wearable electronics and photonics*. Cambridge: Woodhead Publishing in Textiles, pp. 198–220.

[28] Vesely, M., Molenaar, P, Vos, M., Li., R., & Zeiler, W. (2017). Personalized heating – commparison of heaters and control models. *Build Environ*, 112, 223–232.

[29] Orth, M. A., & Berzowska, J. M. (2003). *Electronically controllable, visually dynamic textile, fabric, or flexible substrate. United States Patent Application Publication US 2003/0224155 A1. 2003-12-04.*

[30] Lee, J. A. et al., (2017). Electrochemically powered, energy-conserving carbon nanotube artificial muscles. *Adv Mater*, 29, 1–7, doi: 10.1002/adma.201700870.

[31] Varnaitė, S., & Katunskis. (2009). Influence of washing on the electric charge decay of fabrics with conductive yarns. *Fibres Text East Eur*, 17, 69–75.

[32] Čunko, R., & Andrassy, M. (2005). *Vlakna*. Zagreb: Zrinski, p.71.

[33] Clevertex. Development of a strategic Master Plan for the transformation of the traditional textile and clothing into a knowledge driven industrial sector by 2015. *Report on intelligent textiles. State of the art*. Available from: http://109.2.243.201/documents/State_of_the_art.pdf. Accessed: August 17, 2017.

[34] Suh, M. (2010). E-textiles for wearability: review of integration technologies. Available from: http://www.textileworld.com/textile-world/features/2010/04/e-textiles-for-wearability-review-of-integration-technologies. Accessed: August 19, 2017.

[35] Atkinson, K. R. (2007). Multifunctional carbon nanotube yarns and transparent sheets: fabrication, properties, and applications. *Phys B Condens Matter*, 394, 339–343.

[36] Hunt, C., Ashayer-Soltani, R., & Wills, K. (2015). Smart textiles for wearable technology. Available from: http://www.npl.co.uk/

upload/pdf/201509-smart-textiles-for-wearable-technology-presentation.pdf. Accsessed: August 19, 2017.

[37] Han, P., Zhang, X., & Qiao, J. (2016). Intrinsically conductive polymer fibers from thermoplastic trans-1,4-polyisoprene. *Langmuir*, 32, 4904–4908, doi: 10.1021/acs.langmuir.6b01333.

[38] Maity, S., Chatterjee, A., Singh, B., & Singh, A. P. (2013). Polypyrrole based electro-conductive textiles for heat generation. *J Text I*, 105, 887–893, doi: 10.1080/00405000.2013.861149.

[39] Chatterjee, A., Manivannan, B., & Maity, S. (2017). Electro-conductive Palmyra fibers by in situ polymerisation of pyrrole. *J Nat Fibers*, 14, 185–195, doi: 10.1080/15440478.2016.1193082.

[40] Lim, T. H., Lee, S. H. & Yeo, S. Y. (2017). Higly conductive polymer/metal/carbon nanotube composite fiber prepared by the melt-spinning process. *Tex Res J*, 87, 593–606.

[41] Meoli, D., & May-Plumlee, T. (2002). Interactive electronic textile development: A Review of Technologies. *JTATM*, 2, 1–12.

[42] Stoppa, M., & Chiolerio, A. (2014). Wearable electronics and smart textiles: a critical review. *Sensors*, 14, 11957–11992.

[43] Kamyshny, A., & Magdassi, S. (2014). Conductive nanomaterials for printed electronics. *Small*, 10, 3515–3535, doi: 10.1002/smll.201303000.

[44] Cheng, T. S. (2015). *Printing conductive tracks on textile material. Project report*. Faculty of Mechanical and Manufacturing Engineering, University Tun Onn Malaysia.

[45] Printed copper inks now available for conductive patterns. Available from: http://www.machinedesign.com/materials/interview-printed-copper-inks-now-available-conductive-patterns. Accessed: August 7, 2017.

[46] Fuller, S. B., Wilhelm, E. J., & Jacobson J. M. (2012). Ink-jet printed nanoparticle microelectromechanical systems. *J Microelectromech S*, 11, 54–60, doi: 10.1109/84.982863.

[47] Walker, S. B., & Lewis, J. A. (2012). Reactive silver inks for patterning high-conductivity features at mild temperatures. *J Am Chem Soc*, 134, 1419–1421, doi: 10.1021/ja209267c.

[48] Secor, E. B., & Hersam, M. C. Graphene inks for printed electronics. Available from: www.sigmaaldrich.com/technical-documents/ articles/technology-spotlights/graphene-inks-for-printed-electronics.html. Accessed: August 7, 2017.

[49] Wang, Z., Wang, W., Jiang, Z., & Yu, D. (2016). Low temperature sintering nano-silver conductive ink printed on cotton fabric as printed electronics. *Prog Org Coat*, 10, 604–611. doi.org/ 10.1016/j.porgcoat.2016.08.019.

[50] Jolke, P., Mark, K., Hendriks, C. E., & Schubert, U. S. (2009). Microwave flash sintering of ink jet-printers silver tracks on polymer substrate. *Adv Mater*, 21, 4830–4834, doi: 10.1002/adma.200901081.

[51] Hummeig, M., Zhang, R., Nilsson, H. E., & Olin, H. K. (2011). Electrical sintering of silver nanoparticle ink studies by in situ TEM probing. *PLoS One*, 6, 17209, doi: 10.1371/journal.pone.0017209.

[52] Jeong, S., Song, H. C., Lee, W. W., Choi, Y., & Ryu, V. (2010). Preparation of aqueous Ag ink with long term dispersion stability and its ink jet printing for fabrication conductive tracks on polyamide film. *J Appl Phys*, 108, 102085, doi: 10.1063/1.3511686.

[53] Magdassi, S., Grouchko, M., Berezin, O., & Kamyshny, A. (2010). Triggering the sintering of silver nanoparticles at room temperature. *ACS Nano*, 4, 1943–1948, doi: 10.1021/nn901868t.

[54] Grouchko, M., Kamyshny, A., Mihailescu, C. F., Anghel D. F., & Magdassi, S. (2011). Conductive inks with a built-in mechanism that enables sintering at room temperature. *ACS Nano*, 5, 3354–3359, doi: 10.1021/nn2005848.

[55] Printed electronic inks and coatings - Loctite, Henkel Electronic Materials. Available from: http://www.loctite.sg/sea/content_data/ 405381_Printed_Electronics_Inks_and_Coatings.pdf. Accessed: August 7, 2017.

[56] Graphene inks: introduction and market status. Available from: https://www.graphene-info.com/graphene-inks. Accessed: August 7, 2017.

[57] Skrzetuska, E., Puchalski, M., & Krucinska, I. (2014). Chemically driven printed textile sensors based on graphene in carbon nanotubes. *Sensors*, 14, 16816–16828, doi: 10.3390/s140916816.

[58] Suh, M. (2011). *Development of wireless transmissions between inductively coupled layers in smart clothing: dissertation*. Raleigh: North Carolina State University. Available from: https://repository.lib.ncsu.edu/bitstream/handle/1840.16/7758/etd.pdf?sequence=1&isAllowed=y. Accessed: August 17, 2017.

[59] Kipphan, H. (2001). *Handbook of print media technologies and production methods*. Edited by Kipphan, H. Berlin, Heidenberg, New York: Springer-Verlag.

[60] Kastner, J., et al., (2017). Silver based reactive inks for inkjet-printing of conductive lines. *Microelectron Eng*, 176, 84–88.

[61] Suh, M., Carolla, K. E., Granth, E., & Oxenhama, W. (2012). Effect of fabric substrate and coating material on the quality of conductive printing. *J Text I*, 104, 213–222. doi.org/10.1080/00405000.2012.714107.

[62] WRAP. (2016) Textiles market situation report WRAP: Banbury, UK, Spring 2016, 1–19 Available from: http://www.wrap.org.uk/sites/files/wrap/Textiles_Market_Situation_Report_2016.pdf. Accessed: August 17, 2017.

[63] YnFx. (2016). Steady growth in technical textile visible for global textile companies. Available from: http://www.yarnsandfibers.com/news/news-tags/global-textile-consumption. Accessed: August 17, 2017.

[64] Furlong, H. (2016). Trending: Levi's, EU project Unlock new ways to turn textile waste into new products. Sustainable brands. Available from: http://www.sustainablebrands.com/news_and_views/next_economy/hannah_furlong/trending_levis_eu_project_unlock_new_ways_turn_textile. Accessed: August 17, 2017.

[65] Zacune, J. (2013). Less is more. Resource efficiency through waste collection, recycling and reuse of aluminium, cotton and lithium in Europe. February 2013, 1–30. Available from: https://www.

foeeurope.org/sites/default/files/publications/foee_report__less_is_more_0.pdf. Accessed: August 17, 2017.

[66] Council for textile recycling. (2017). Available from: http://www.weardonaterecycle.org/. Accessed: August 17, 2017.

[67] Cuc, S., & Vidovic, M. (2011). Environmental sustainability through clothing recycling. *OSCM*, 4, 108–115.

[68] Innovation in textiles. (2016). Global e-textiles market to grow with CAGR of around 36.2%. Available from: http://www.innovationintextiles.com/global-etextiles-market-to-grow-with-cagr-of-around-362/. Accessed: August 17, 2017.

[69] Russell, M. (2016). E-textiles market set to reach US$3bn by 2026. Available from: https://www.just-style.com/news/e-textiles-market-set-to-reach-us3bn-by-2026_id126932.aspx. Accessed: August 17, 2017.

[70] Köhler, A. R., Hilty, L. M., & Bakker, C. (2011). Prospective impacts of electronic textiles on recycling and disposal. *J Ind Ecol*, 14, 496–510, doi: 10.1111/j.1530-9290.2011.00358.x.

[71] Köhler, A. R. (2013). Challenges for eco-design of emerging technologies: the case of electronic textiles. *Mater Des*, 51, 51–60, doi:10.1016/j.matdes.2013.04.012.

[72] Hansen, A-L. D., Hansen, M., & Jensen, H. T. (2011). *Future textiles trends inspiration drivers & barriers smart textiles*. Herning: Knowledge Centre for Smart Textiles.

[73] Karlsson, R. & Luttropp, C. (2006). Eco design: what's happening? An overview of the subject area of eco design and of the papers in this special issue. *J Cleaner Prod*, 14, 1291–1298.

[74] Rossi, D. D., & Paradiso, R. (2011). Future direction: e-textiles. In: *Wearable monitoring systems*. Edited by Bonfiglio, A. & Rossi, D. D. New York: Springer US, 147–162.

[75] Van den Velden, N. M., Kuusk, K., & Köhler, A. R. (2015). Life cycle assessment and eco-design of smart textiles: the importance of material selection demonstrated through e-textile product redesign. *Mater Des*, 84, 313–324, doi: 10.1016/j.matdes.2015.06.129.

[76] European Commission. (2008). Directive 2008/98/EC of the European Parliament and of the Council on waste and repealing certain Directives. *Off J Eur Union*, L 312/3, November 22, pp. 3–30.
[77] European Commission. (2002). Directive 2002/95/EC of the European Parliament and of the council on the restriction of the use of certain hazardous substances in electrical and electronic equipment. *Off J Eur Union,* L 037, February 13, 2002, pp. 0019–0023.
[78] European Commission. (2012). Directive 2012/19/EU of the European Parliament and of the Council on waste electrical and electronic equipment (WEEE). *Off J Eur Union*, L 197/38, July 24, 2012, pp. 38–71.
[79] European Commission. (2006). Directive 2006/66/EC of the European Parliament and of the Council on batteries and accumulators and waste batteries and accumulators and repealing Directive 91/157/EEC. *Off J Eur Union*, L 266, September 26, 2006, pp. 1–14.
[80] Tiberto, P., Barrera, G., Celegato, F., Coisson, M., Chiolerio, A., Martino, P., Pandolfi, P., & Allia, P. (2013). Magnetic properties of jet-printer inks containing dispersed magnetite nanoparticles. *EPJ B*, 83,173–179, doi: 10.1140/epjb/e2013-30983-8.
[81] Harbisson, N. (2012). I listen to color. TED Global. Available from: http://www.ted.com/talks/neil_harbisson_i_listen_to_color.html. Accessed: July 24, 2017.
[82] Schiphorst, T. (2011). Self-evidence: applying somatic connoisseurship to experience design. *Proceedings of the International conference on human factors in computing systems*, CHI 2011, Vancouver, Canada, 145–160.
[83] Quinn, B. (2013). *Textile visionaries: innovation and sustainability in textiles design*. London: Laurence King Publishing, 312.
[84] Flanagan, P., Papadopoulos, D., & Voss, G. (2016). Intimacy and intimacy: ethics, power and potential of wearable technologies. In: *Fundamentals of wearable computers and augmented reality. 2nd ed.*

Edited by Woodrow Barfield. Boca Raton: CRC Press, pp. 31–56, doi: 10.1201/b18703-5.

[85] Reiss, A., & Amft, O. (2016). Design challenges of real wearable computers. In: *Fundamentals of wearable computers and augumented reality. 2nd ed.* Edited by Woodrow Barfield. Boca Raton: CRC Press, pp. 583-619, doi: 10.1201/b18703-27.

[86] Quintana, M., Wander, S., Wu, A., & Brigante, M. (2014). Coloring, future wearables. Project coloring. *MFA Interaction Design.* Available from: http://interactiondesign.sva.edu/people/project/coloring. Accessed: July 26, 2017.

[87] Confino, J. (2013). Google seeks out wisdom of Zen master Thich Nhat Hanh. *The Guardian,* September 5, 2013. Available from: http://www.theguardian.com/sustainablebusiness/global-technology-ceos-wisdom-zen-master-thich-nhat-hanh. Accessed: July 26, 2017.

[88] Meoli, D. (2002). *Interactive electronic textiles: technologies, applications, opportunities, and market potential: MS.c. Thesis.* Rayleigh: North Carolina State University.

[89] Unitika LTD. Available from: http://www.unitika.co.jp/e. Accessed: July 24, 2017.

[90] Karaguzel, B., Merritt, C. R., Kang, T., Wilson, J. M., Nagle, H. T., Grant, E., & Pourdeyhimi, B. (2009). Flexible, durable printed electrical circuits. *J Text I,* 100, 1–9.

[91] Pold, S. (2005). Interface realisms: the interface as aesthetic form. *PMC,* Vol. 15, doi: 10.1353/pmc.2005.0013.

[92] iQ, Intel, PSFK Report. (2014). Available from: https://iq.intel.com/. Accessed: July 28, 2017.

[93] PSFK Labs. (2014). The future of wearable tech. Available from: https://www.slideshare.net/PSFK/psfk-future-of-wearable-technology-report. Accessed: July 24, 2017.

Biographical Sketch

Maja Klančnik, PhD

Affiliation: University of Ljubljana, Faculty of Natural Sciences and Engineering, Department of Textiles, Graphic Arts and Design

Education: PhD

Business Address: Snežniška 5, SI- Ljubljana, Slovenia

Research and Professional Experience: Environmental impact of printing and textile industry, Treatment of wastewaters, Printing forms and printing plates, Dyeing of textile fibers, Kinetics of reactions of reactive dyes, Printed wearable electronics

Professional Appointments: Assistant Professor

Publications from the Last 3 Years:

Vrabič Brodnjak, Urška, Gregor-Svetec, Diana, Klančnik, Maja. Influence of enzymatic treatment on the structural, sorption and dyeing properties of viscose and chitosan/cellulose fibers. *Textile research journal*, ISSN 0040-5175, 2016, vol. 86, no. 9, 990-1005.

Klančnik, Maja, Batista, Meta. Čiščenje odpadne vode, onesnažene s sitotiskarsko barvo = Treatment of wastewater contaminated with screen printing ink. *Tekstilec: glasilo slovenskih tekstilcev*, ISSN 0351-3386. 2015, vol. 58, no. 3, 209-220.

Klančnik, Maja. Coagulation and adsorption treatment of printing ink wastewater. *Acta graphica : revija za grafičku tehnologiju, inžinjerstvo i dizajn*, ISSN 0353-4707. 2014, vol. 25, no. 3/4, 73-82.

Jereb, Ana, Javoršek, Dejana, Klančnik, Maja. Comparision of two thermal decomposition offset printing plates. *Acta graphica : revija za grafičku tehnologiju, inžinjerstvo i dizajn*, ISSN 0353-4707. 2014, vol. 25, no. 3/4, 91-100.

Velechovsky, Tina, Urbas, Raša, Klančnik, Maja. Application of conventionally made cliche for pad transfer printing on nails. In: Urbas, Raša (ed.). *Proceedings*, 7th Symposium of Information and Graphic Arts Technology, Ljubljana, 5-6 June 2014. Ljubljana: Faculty of Natural Sciences and Engineering, Department of Textiles. 2014, pp. 211-217.

Tabor, Tanja, Klančnik, Maja. Adsorption of printing ink from wastewater. In: Urbas, Raša (ed.). *Proceedings*, 7th Symposium of Information and Graphic Arts Technology, Ljubljana, 5-6 June 2014. Ljubljana: Faculty of Natural Sciences and Engineering, Department of Textiles. 2014, pp. 235-240.

Marica Staresinic PhD

Affiliation: University of Ljubljana, Faculty of Natural Sciences and Engineering, Department of Textiles, Graphic Arts and Design

Education: PhD in information technologies

Business Address: Snežniska 5, 1000 Ljubljana SLOVENIA

Research and Professional Experience: printed electronics, image analysis, interactive systems, materials analysis, Photography - *processing and reproduction of data.*

Professional Appointments: Assistant Professor

Publications from the Last 3 Years:

1. Starešinič, Marica, Ahtik, Jure, Učakar, Andrej, Javoršek, Dejana. Evaluation of changes in urban environments using orthophoto maps. V: Cohen, Alexis (ur.). *Urban and built environments: sustainable development, health implications and challenges*,

(Urban development and infrastructure). New York: Nova Science Publishers. 2015, str. [41]-53, ilustr. [COBISS.SI-ID 3187056].
2. Javoršek, Dejana, Močnik, Janja, Starešinič, Marica. Analyses of colour appearances on different display devices = Analiza interpretacije barv na različnih prikazovalnih napravah. *Tekstilec: glasilo slovenskih tekstilcev*, ISSN 0351-3386. [Tiskana izd.], 2015, letn. 58, št. 2, str. 100-107, ilustr. [COBISS.SI-ID 3139440]
3. Starešinič, Marica, Javoršek, Andrej, Javoršek, Dejana. Textiles using electronic applications. V: Mondal, Ibrahim H. (ur.). *Textiles: history, properties and performance and applications*, (Chemistry research and applications). New York: Nova Science Publishers. 2014, str. [447]-463, ilustr. [COBISS.SI-ID 3002992].
4. Ahtik, Jure, Muck, Deja, Starešinič, Marica. A novel database for evaluation of digital images. V: URBAS, Raša (ur.). *Abstracts*, 7th Symposium of Information and Graphic Arts Technology, Ljubljana, 5-6 June 2014. Ljubljana: Faculty of Natural Sciences and Engineering, Department of Textiles, Chair of Information and Graphic Art Technology. 2014, str. 49. [COBISS.SI-ID 3043184]
5. Starešinič, Marica, Javoršek, Andrej, Javoršek, Dejana. Visokotehnološka oblačila - kaj bomo nosili v prihodnosti? = Hightech dress - what shall we wear in the future?. V: Simončič, Barbara (ur.), Forte-Tavčer, Petra (ur.). *Trajnostni razvoj v tekstilstvu: zbornik izvlečkov*, 44. simpozij o novostih v tekstilstvu, Ljubljana, 6. junij 2013. Ljubljana: Naravoslovnotehniška fakulteta, Oddelek za tekstilstvo. 2013, str. 48. [COBISS.SI-ID 2880880]

Helena Gabrijelčič Tomc. PhD

Affiliation: University of Ljubljana

Education: PhD

Business Address: Faculty of Natural Sciences and Engineering, Department of Textiles, Graphic Arts and Design, Snežniška 5, 1000 Ljubljana, Slovenia

Research and Professional Experience:
1. A development of *constructional and color-mixing models* based on physical products and computer generated simulations; 2. generating, processing and *reproduction of data in media* (shape, pattern, color) and three-dimensional computer graphic and visualizations; 3. Research and developing the methodologies of teaching and learning approaches for web site planning, development and testing and implementation of 2D and 3D computer graphic in creative processes of studying.

Professional Appointments: Associate Professor

Publications from the Last 3 Years:

Bratuž, Nika, Javoršek, Dejana, Gabrijelčič TOMC, Helena. Defining optimal conditions of colors in 3D space in dependence on gamma values, illumination, and background color. *Journal of imaging science and technology*, ISSN 1062-3701, Jul./Aug. 2015, vol. 59, no. 4, pp. 40503/1-40503/11

Gabrijelčič Tomc, Helena, Hladnik, Aleš. 1D and 2D shape descriptors applied in fabric drape computer simulation. *Fibres & textiles in Eastern Europe: an international magazine devoted to current problems of the textile industries in Central and Eastern Europe*, ISSN 1230-3666, 2015, vol. 23, no. 6 (114), pp. 92-101

Arandjus, Mark, Gabrijelčič Tomc, Helena. Rendering 3D graphics as an aid to stylized line drawings in perspective. *JGED: Journal of Graphic Engineering and Design*, ISSN 2217-379X, 2016, vol. 7, no. 2, pp. 5-13.

Iskra, Andrej, Gabrijelčič Tomc, Helena. Eye-tracking analysis of face observing and face recognition. *JGED: Journal of Graphic Engineering and Design*, ISSN 2217-379X, 2016, vol. 7, no. 1, pp. 5-11.

Varga, Luka, Gabrijelčič Tomc, Helena. Subliminal messages in an online environment. *Acta graphica: revija za grafičku tehnologiju, inžinjerstvo i dizajn*, ISSN 0353-4707. [Print ed.], 2016, vol. 27, no. 4, pp. 23-32

Bratuž, Nika, Gabrijelčič Tomc, Helena, Javoršek, Dejana. Rendering techniques in 3D computer graphics based on changes in the brightness of the object background. V: Cvetković, Dragan (ur.). *Computer simulation*, (Computer and Information Science, Computer Science and Engineering). Rijeka: InTech. 2017, pp. [21]-44.

Kočevar, Tanja Nuša, Gabrijelčič TOMC, Helena. Modelling and visualisation of the optical properties of cloth. V: CVETKOVIĆ, Dragan (ur.). *Computer simulation*, (Computer and Information Science, Computer Science and Engineering). Rijeka: InTech. 2017, pp. [45]-65

Diana Gregor-Svetec, PhD

Affiliation: University of Ljubljana, Faculty of Natural Sciences and Engineering

Education: PhD

Business Address: Department of Textiles, Graphic Arts and Design, Snezniska 5, SI-1000 Ljubljana, Slovenia

Research and Professional Experience: characterization of textile and graphic materials, printing substrates and packaging materials, recycling, eco-design of printed products and sustainable packaging

Professional Appointments: Full Professor

Publications from the Last 3 Years:

Vrabič Brodnjak, Urška, Gregor-Svetec, Diana, Klančnik, Maja. Influence of enzymatic treatment on the structural, sorption and dyeing properties of viscose and chitosan/cellulose fibers. *Textile research journal*, ISSN 0040-5175, 2016, vol. 86, no. 9, pp. 990-1005, doi: 10.1177/0040517514559577.

Rožić, Mirela, Kulčar, Rahela, Jamnicki, Sonja, Lozo, Branka, Gregor-Svetec, Diana. UV stability of thermochromic ink on paper containing clinoptilolite tuff as a filler. *Cellulose chemistry and technology*, ISSN 0576-9787, 2015, vol. 49, no. 7/8, pp. 693-699.

Gregor-Svetec, Diana, Pivar, Matej, Muck, Deja. Properties of recycled papers important for printable antennas. *Cellulose chemistry and technology*, ISSN 0576-9787, 2015, vol. 49, no. 7/8, pp. 701-708.

Kavčič, Urška, Pivar, Matej, Đokić, Miloje, Gregor-Svetec, Diana, Pavlovič, Leon, Muck, Deja. UHF RFID tags with printed antennas on recycled papers and cardboards. *Materiali in tehnologije*, ISSN 1580-2949, 2014, vol. 48, no. 2, pp. 261-267.

Tišler Korljan, Barbara, Gregor-Svetec, Diana. Properties and printability of compression moulded recycled polyethylene. *Materials & design*, ISSN 0264-1275, Mar. 2014, vol. 55, pp. 583-590.

Zule, Janja, Černič, Marjeta, Gregor-Svetec, Diana. Recyclability determination of paper based packaging products. *Papir: revija Društva inženirjev in tehnikov papirništva*, ISSN 0350-6614, 2014, vol. 42, no. 11, pp. 36-40.

Vodopivec, Jedert, Gregor-Svetec, Diana, König, Silva, Černič, Marjeta. Papir v knjižnem bloku Slave vojvodine Kranjske iz leta 1689 = Paper in the text block of the book The Glory of the Dutchy of Carniola. *Papir : revija Društva inženirjev in tehnikov papirništva*, ISSN 0350-6614, 2014, vol. 42, no. 12, pp. 37-41.

Gregor-Svetec, Diana, Zule, Janja, Elegir, Graziano. Recyclability of graphic products and paper based packaging. V: *Wood, pulp & paper 2014: [proceedings of joint conference = zbornik vedeckých konferencií]*, Wood, Pulp and paper 2014 and Polygrafia academica

2014, Bratislava 12.-13. March 2014. Bratislava: Slovenská technická univerzita v Bratislave. 2014, pp. 14-19.

Golja, Viviana, Gregor-Svetec, Diana, Pivar, Matej, Zorič, Andreja, Novak, Saša. Detection of silver nanoparticles present in food contact paper with simple screening method. V: *Matbim 2017, 4th International Meeting on Material/Bioproduct Interaction, Porto, Portugal, 26th-28th April 2017: book of abstracts*. Porto. 2017, str. 86-88.

Gregor-Svetec, Diana, Muck, Deja, Pivar, Matej, Ravnjak, David, Bolanča-Mirković, Ivana. Recyclability of label papers with integrated UHF RFID printed antenna. In: Kleebauer, Markus (ed.), Gericke, Ralf (ed.). *Proceedings: joint conference of PTS and COST Action*, PTS Innovative Packaging Symposium 2016, April 6th - 7th, 2016, Munich. Munich: PTS. 2016, pp. 127-140.

Heršold, Špela, Gregor-Svetec, Diana. Reusable eco carrier bag from bamboo. In: Simončič, Barbara (ed.), TOMŠIČ, Brigita (ed.), Gorjanc, Marija (ed.). *Proceedings*, 16th World Textile Conference Autex 2016, 8-10 June 2016, Ljubljana, Slovenia. Ljubljana: Faculty of Natural Sciences and Engineering, 2016, 6 p.

Gregor-Svetec, Diana, Muck, Deja, Pivar, Matej, Đokić, Miloje, Kavčič, Urška. Printed electronics on recycled paper, cardboard and foils. In: *Existing technologies and current developments in active and intelligent packaging*, COST Action FP1405 Meeting, Portugal, Aveiro, September 15 -16, 2015. 13 p.

Gregor-Svetec, Diana, Muck, Deja, Pivar, Matej, Golob, Gorazd, Urbas, Raša, Szentgyörgyvölgyi, Rozália, Kavčič, Urška. Novelties in use of recycled paper and paper based packaging. In: Borbély, Ákos (ed.). *Conference proceedings*, International Conference on Design and Light Industry Technologies, November 19-20, 2014, Budapest, Hungary. Budapest: Óbuda University, Faculty of Light Industry and Environmental Engineering. 2014, pp. 23-32.

Gregor-Svetec, Diana, Zule, Janja, Putz, Hans-Joachim, Elegir, Graziano. Recyclability evaluation of paper based packaging. In: URBAS, Raša (ed.). *Proceedings*, 7th Symposium of Information and Graphic Arts Technology, Ljubljana, 5-6 June 2014, 2014, pp. 12-17.

Gregor-Svetec, Diana, Možina, Klemen, Sežun, Mija, Grossmann, Harald. Recycling and waste management in Central Europe countries. In: URBAS, Raša (ed.). *Proceedings*, 7th Symposium of Information and Graphic Arts Technology, Ljubljana, 5-6 June 2014, 2014, pp. 18-23.

Golob, Gorazd, Gregor-Svetec, Diana, Urbas, Raša, Mešl, Mateja, Zule, Janja, GROß, Anja. Assessment of possible improvement of EU legislation on paper recyclability. In: Urbas, Raša (ed.). *Proceedings*, 7th Symposium of Information and Graphic Arts Technology, Ljubljana, 5-6 June 2014, 2014, pp. 37-41.

Gregor-Svetec, Diana, Arko, Manca, Možina, Klemen. Graphic paper from eucalyptus. In: URBAS, Raša (ed.). *Proceedings*, 7th Symposium of Information and Graphic Arts Technology, Ljubljana, 5-6 June 2014, 2014, pp. 218-222.

Tatjana Rijavec, PhD

Affiliation: University of Ljubljana, Faculty of Natural Sciences and Engineering, Department of Textiles, Graphic Arts and Design

Education: PhD, textile technology

Business Address: Snežniška 5, 1000 Ljubljana, Slovenia

Research and Professional Experience: structure and properties of textile and high performance fibres; smart and nanomaterials (nitinol, silica aerogel)

Professional Appointments: editor-in-chief of Slovenian professional journal Tekstilec (www.tekstilec .si) from 2012 onwards

Publications from the Last 3 Years:

Rijavec, Tatjana, Jevšnik, Simona, Soljačič, Ivo, Tomljenović, Antoneta, VIŠIĆ, Ksenija, Pušić, Tanja. Physical and chemical effects of washing and wet cleaning on durable press finished cotton fabrics. *Textile research journal*, ISSN 0040-5175, 2015, vol. 85, no. 6, 632-645, doi: 10.1177/0040517514547212.

Kraner Zrim, Polona, Mekjavić, Igor B., Rijavec, Tatjana. Properties of laminated silica aerogel fibrous matting composites for footwear applications. *Textile research journal*, ISSN 0040-5175, 2016, vol. 86, no. 10, 1063-1073, doi: 10.1177/0040517515591781.

Alej, Alenka, Fajfar, Peter, Lavrič, Zoran, Bukošek, Vili, RIJAVEC, Tatjana. Preparation of shape memory NiTiNOL filaments for smart textiles. *Tekstilec: glasilo slovenskih tekstilcev*, 2016, 59(2), 168-174, doi: 10.14502/Tekstilec2016.59.168-174.

Rijavec, Tatjana, Janjić, Svjetlana, KOCJAN Ačko, Darja. Revitalization of industrial hemp Cannabis sativa L. var. sativa in Slovenia: a study of green hemp fibres. *Tekstilec,* 2017, 60(1), 36-48, doi: 10.14502Tekstilec2017.60.36-48.

In: Printed Electronics
Editors: K. Yvon and N. Fabrice
ISBN: 978-1-53612-797-3
© 2017 Nova Science Publishers, Inc.

Chapter 3

DESIGNING WEARABLE SYSTEMS-ON-POLYMER USING FLEXIBLE HYBRID ELECTRONICS

Umit Y. Ogras,* Ujjwal Gupta, Jaehyun Park and Ganapati Bhat
School of Electrical, Computer, and Energy Engineering
Arizona State University, Tempe, AZ, US

Abstract

Flexible electronics have the potential to revolutionize the way humans interact, communicate, and control electronic devices. Cost and weight advantages combined with physical flexibility open a wide range of application areas, including wearable electronics, medical devices, and internet of things (IoT). However, performance and capabilities offered by flexible electronics technology are severely limited, and far lower than traditional rigid systems-on-chip (SoC) used today. Emerging flexible hybrid electronics (FHE) target this problem by integrating traditional rigid ICs and printed electronics on a flexible substrate, such as polymer. FHE can be used to coalesce rigid and flexible resources judiciously to drive the next big leap forward in the form factor design, similar to the shift from desktop and laptop computers to hand-held devices. This chapter compares the FHE systems-on-polymer (SoPs) with traditional SoCs and

*Corresponding Author Email: umit@asu.edu

introduces the physical flexibility as a new dimension in the design space. It overviews the major requirements of flexibility-aware design methodologies and tools tailored for FHE. Finally, it discusses near- and long-term application areas and relevant challenges.

Keywords: flexible hybrid electronics, systems-on-polymer, flexibility-aware system optimization, wearable computing, energy harvesting

1. Introduction

Flexible electronics technology has the potential to transform computing by enabling bendable and stretchable wearable systems at a low cost [1]. Physical flexibility combined with cost and weight advantages opens a wide range of form factors and application areas, including wearable electronics, medical sensing, rollable displays, and Internet of Things (IoT) [2–5].

While impressive progress has been made in recent years, flexible electronics still have significantly lower performance [6, 7] and larger parameter variations [8] compared to silicon CMOS circuits. For example, silicon technology offers 14 nm feature size with over 1 GHz frequency, whereas the feature sizes of thin-film transistors (TFT) range from 8 μm to 50 μm [9], and frequencies hardly reach 10 MHz [10]. Therefore, practical use of flexible electronics is largely limited to displays [11], sensors [12], photovoltaic cells [13] and wireless tags [14]. More recent examples include batteries [15, 16], programmable logic circuits [17], simple micro-controllers, a flexible system-on-chip, analog-to-digital converters (ADC) and radio frequency transmitters [18, 19] [1]. Stretchable electronics are more general than flexible circuits in the sense that they can stretch as well as bend. Recent work demonstrated metal interconnects for stretchable electronics [20, 21]. For example, a stretchable horseshoe shaped interconnects are demonstrated in [20]. However, performance and capabilities of purely flexible electronics are still limited compared to traditional electronics [7, 8].

Flexible hybrid electronics (FHE) technology has recently been introduced to combine the advantages of flexible and traditional Silicon technolo-

[1] A comprehensive overview of flexible electronics with historical perspective and associated design challenges are presented by Huang et al. [6].

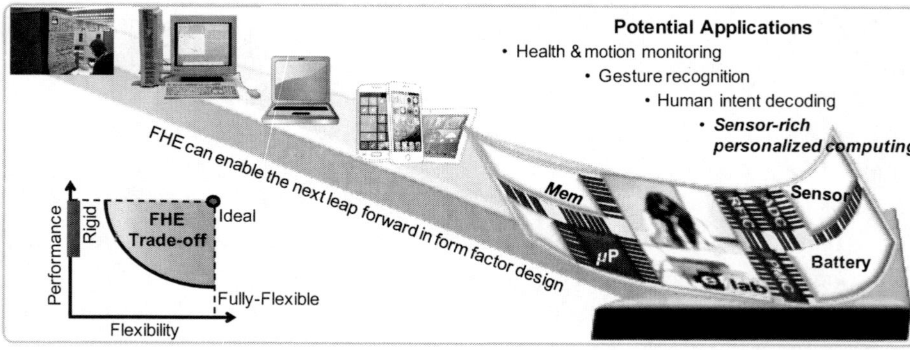

Figure 1. Evolution of the form factor, and the System-on-Polymer (SoP) proposal with flexible circuits, display, battery and rigid integrated circuits (ICs).

gies [22, 23]. FHE can be used to coalesce rigid and flexible resources judiciously to bridge the gap between today's complex systems and the advantages of flexible electronics. Integration with the CMOS technology enables sensing, processing and communication in a truly wearable form factor. Furthermore, FHE devices can have arbitrary shapes to surround objects and conform to the body, while incorporating multiple functions. As a result, FHE can help in transforming personalized computing by providing a systematic approach to design wearable systems and arbitrarily shaped objects. For example, FHE can enable electronic patches equipped with motion, physiological and biochemical sensors, and wireless transceivers. These patches can be used to examine movement disorders *anytime* and *anywhere*, unlike the current practice which requires patient to stay in a clinical environment [24, 25]. Promising near-term applications of the FHE include assistive IoT technologies [26, 27], seamless health monitoring, motion processing, gesture recognition and soft-robotics for elderly. In the longer term, FHE can deliver a richer experience than current smartphones in a truly pervasive form factor with the help of (1) physical flexibility (e.g., rollable or foldable devices become possible) and (2) wearable sensors that enable machines to understand and assist their users. Hence, FHE can drive the next big leap forward in the form factor design, similar to the shift from desktop and laptop computers to hand-held devices, as illustrated in Figure 1.

FHE technology is recognized as a promising technology that can combine the physical benefits of flexible electronics with the computational power of CMOS technology [5, 7, 28]. Recently established FHE Manufacturing Innovation Institute [2] takes a step further and addresses the industry-relevant problems. Integration of CMOS devices on flexible substrates has recently been demonstrated at research centers including ASU Flexible Display Center [7,29], industry [18] and academia [30]. Moreover, circuits for interfacing flexible electronics and CMOS ICs have been proposed [31]. These early studies are critical milestones towards enabling FHE systems. However, there are no known design methodologies for flexible hybrid electronic systems [2]. Therefore, there is a need for theoretically grounded and practically applicable tools that quantify physical flexibility as a new metric, and incorporate it in the design process. Towards this end, the rest of this chapter is organized to overview the following areas:

- Introducing physical flexibility as a new dimension in the design space, in addition to traditional metrics, such as performance, power consumption and reliability,
- Constructing new methodology and tools for FHE similar to those for system-on-chips (SoCs),
- Innovating practical energy harvesting and runtime management techniques tailored for FHE systems, which have no or very limited battery power,
- Developing applications and FHE prototypes that can drive new design methodologies, while demonstrating the potential of the FHE technology.

2. Flexibility: A New Dimension in the Design Space

Systems based on flexible hybrid electronics differ from traditional computing systems, such as SoCs, in a number of critical ways. The most notable difference is the physical flexibility, which can result in shape changes due to bending and stretching. As a result, physical flexibility and stretchability expands the design space into an uncharted dimension, which introduces intricate tradeoffs with the traditional power, performance and area metrics, as illustrated in Figure 2.

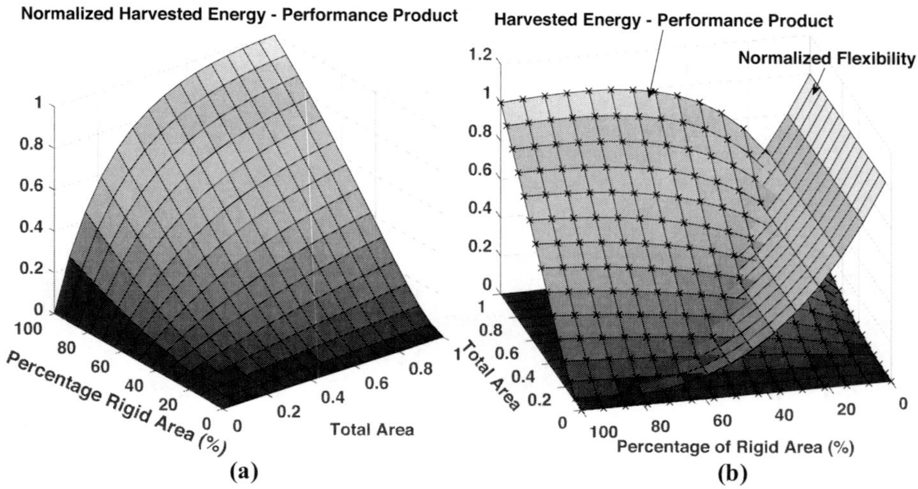

Figure 2. (a) Harvested Energy–Performance product (W×Ops), (b) The trade-off between normalized flexibility and harvested Energy–performance product as a function of total area and percentage of rigid area.

One of the distinctions of SoPs from SoC is the focus on the wearability, which implies that these systems should have light weight and potentially touch the user skin. Light weight requires batteryless operation or allows at most small printed batteries with limited capacity. Therefore, energy harvesting from ambient sources, such as light, body heat and motion, becomes inevitable. Proximity to the user skin places tight constraints on the surface temperature. These constraints are more stringent than that of SoCs, since large heat sinks are prohibitive due to weight and flexibility constraints. Furthermore, flexible hybrid electronics benefit from, or may even require, larger area to reach sufficiently wide sensing area. When flexible solar cells are used for energy harvesting, generated energy grows proportionally with the surface area. In contrast, traditional chip design techniques typically minimize the area to reduce the cost, and feature sizes to improve transistor timing.

To combine the energy harvesting potential (W×s) and performance (Operations/s, i.e., Ops/s), we can use their product (W×Ops) as a metric for FHE systems. Figure 2(a) illustrates that larger total area can be utilized to increase the capacity of a FHE device measured using this metric. Furthermore, allocating more area to rigid resources yields larger rate of return per unit area, since it allows for faster and larger number of transistors. In general, dedicating a larger area to rigid ICs and interconnect boosts the processing power, since flexible electronics exhibit lower performance and scalability [6, 8]. However, larger number of rigid components and wires would increase the rigid area, and significantly deteriorate the flexibility, as illustrated in Figure 2(b). For instance, increasing the rigid area from 10% to 40% can reduce flexibility by about 3× [32]. This leads to a new set of flexibility-aware optimization problems, such as maximizing the performance or harvested energy under flexibility constraints.

Finally, bending and shape changes may affect the electrical characteristics in ways that were not considered before. For instance, consider a printed antenna, which is an indispensable component of a FHE design with wireless communication capability. Stretching the antenna would impact the center frequency and frequency response, while bending can introduce scattering effects, which are not present in a flat antenna. As an example, Figure 3(a) shows that the power received from an inverted-F antenna at the receiver side can decrease by as much as 5dB with bending [32]. This is roughly equivalent to increasing the separation between the transmitter and receiver by 2.5×. Similarly, energy harvesting resources, such as photo-voltaic (PV) cells, will be crucial components of self-sustained wearable SoPs. Suppose that a PV-cell of length L is bent with a radius of curvature R. The flat case corresponds to $R \to \infty$, while a larger L/R ratio implies more bending. Figure 3(b) clearly shows the generated power as a function of L/R for two different elevation angles of the light source. We observe that the generated power drops by almost half as a result of bending [33].

3. Flexibility-Aware SoP Design Methodology

Remarkable progress in integrated circuit technology is enabled to a large extent by standard design flows [34, 35]. For example, SoC design tools and methodologies enable high level specification of the target systems, and automate synthesis, placement, routing and layout generation [36, 37]. Common strategies, such as re-using intellectual property (IP) cores and raising the level of abstraction, enable power and performance optimized chips with over one billion tran-

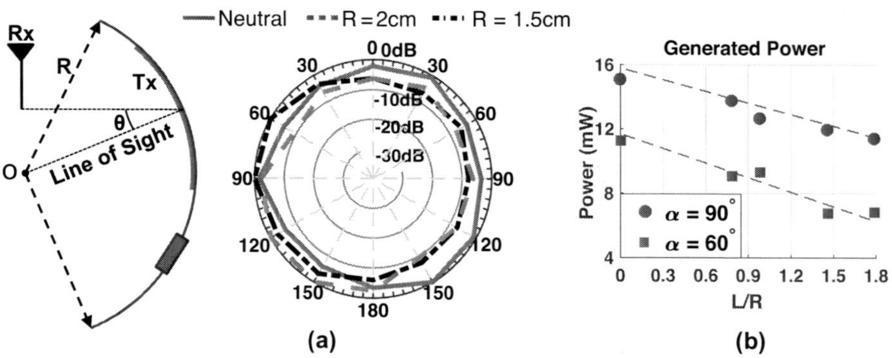

Figure 3. (a) Radiation patterns observed at the receiver when the printed antenna is flat, and bent [32], (b) Power generated by a flexible PV-cell is plotted as a function of degree of bending for two different elevation angles of the light source (α). We observe that the generated power decreases significantly with bending.

sistors under tight time to market constraints [38, 39]. The potential of the FHE technology can be fulfilled by incorporating the newly introduced flexibility dimension to these existing flows and developing new flexibility-aware design tools.

We envision a FHE design flow similar to the one illustrated in Figure 4. The first step towards realizing this flow is quantifying the physical flexibility as a new metric and modeling the impact of flexibility on existing metrics. These models can enable the designers to quantify how the rigid components impact bending and stretching. A component library stores the set of off-the-shelf rigid components, such as microcontrollers, and flexible resources, as depicted in Figure 4. Similar to the use of standard cell libraries, this data can be used by flexibility-aware design tools, which constitute the second major component of the proposed design flow. These tools map a given class of target applications to a subset of the resources in the component library. Selection of the rigid and flexible components impact physical flexibility, as well as traditional power consumption, performance and energy metrics. Since there are multiple ways to implement the required functionality, this leads to a coverage and a constrained optimization problem. The next step is interconnecting the IP

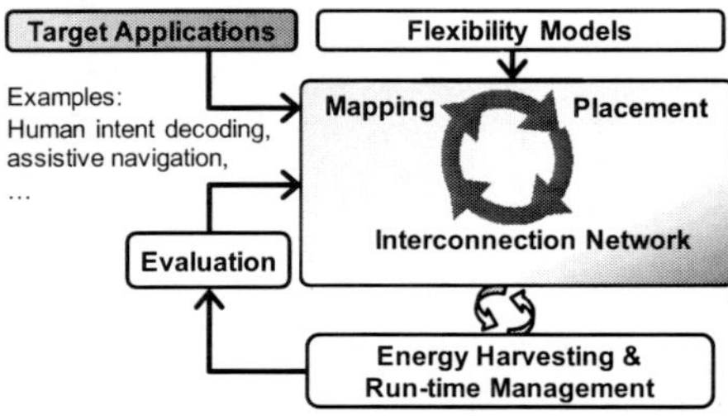

Component Library

Name	Supported Tasks	Size (mm²)	Power (mW)	Throughput (MB/s)	Modulus of Elasticity (Gpa)
Comp-1	CPU process.	19×17	0.92	0.32	0.56
Comp-2	BLE comm	5×5	18.5	0.12	0.56
Comp-3	Gyro & Accel.	3×3	2.31	0.048	270
Comp-4	DSP process. BLE comm.	10×10	2.5 0.98	160	270
...	...	...	...	...	...

Figure 4. An example FHE design flow and sample component library.

cores in the system with an energy efficient and robust interconnection network over flexible and stretchable substrates. Finally, the design time optimization is complemented by energy harvesting management techniques that ensure optimal operation at runtime. The rest of this section discusses these problems in more detail.

3.1. Flexibility-Aware Mapping and Placement

FHE devices integrate a mixture of rigid and flexible IP cores over a flexible or stretchable substrate. Each component has a different degree of flexibility, which can be quantified by the modulus of elasticity [40]. For example, rigid components on the substrate act as pivot points about which the flexible substrate bends, while metal interconnect decrease the level of flexibility. In general, the rigid components may be spread across the entire floorplan of the device. Therefore, the device will have multiple pivot points, and exhibit different deflection behavior along each bending axis. It is possible to quantify these effects using finite element method (FEM) simulations or simpler analytical models [41,42]. Then, these models can be used to compute flexibility of an FHE design as a function of the choice and placement of the IP cores from the component library [32].

Given the flexibility models and component library, the first essential design decision is choosing a subset of rigid and flexible resources to implement the target application. Putting more functionality to a few large rigid ICs would maximize the integration density and deliver larger processing capability, since flexible electronics exhibit poor performance and scalability [6, 8]. However, large silicon chips would also undermine the advantages of flexibility. Therefore, it is important to develop novel techniques to map the applications to conventional ICs and flexible circuits under flexibility, power, performance, and temperature constraints. The optimum design is critically dependent upon the rigid and flexible resources that can be employed in the design and the application requirements. Furthermore, the mapping should be co-optimized simultaneously with placement, since the location and size of the rigid components affect the flexibility significantly [32].

We illustrate flexibility-aware design, using a generic energy minimization formulation under flexibility, power consumption, performance and temperature constraints. Suppose that our goal is to implement the target application G_A by mapping its tasks (T) to the component library (map(T)) to minimize the energy consumption subject to flexibility (C1), power (C2), performance (C3) and thermal constraints (C4): [ams gather, title=Equation 1: Optimum mapping and placement, colframe=gray!100!black]

$$
\begin{aligned}
&\text{Find } map(T) \text{ to minimize } Energy \text{ subject to}\\
&\text{C1}: \text{ Flexibility}(map(T), Floorplan) \geq \delta_{\text{min}}\\
&\text{C2}: Power(G_{\text{A}}, map(T)) \leq \mathbf{P}_{\text{Budget}},\\
&\text{C3}: t_{exe}(G_{\text{A}}, map(T)) \leq \text{t}_{\text{max}}, && (1)\\
&\text{C4}: G_t^{-1} \times Power(G_{\text{A}}, map(T)) + T_{\text{amb}} \leq \mathbf{T}_{\text{max}}\\
&\quad \text{where } G_t: \text{ Thermal conductance matrix,} && (2)\\
&\qquad\quad T_{\text{amb}}: \text{Ambient temperature}
\end{aligned}
$$

In this formulation, Flexibility(map(T)) quantifies the degree of flexibility as a function of the IP cores employed in the design and their placement. Constraint C1 ensures that the final design satisfies the flexibility requirement of the target application. Similarly, C2 through C4 express the flexibility-aware power, response time, and temperature constraints. Variations of this formulation can be used for other optimization objectives, such as maximizing the energy efficiency.

The solution to this problem is difficult in general due to the complexity of both mapping and placement problems [43,44]. One preliminary approach to solve the problem of mapping tasks to rigid and flexible components is presented in [45]. This approach first builds a library of rigid and flexible components as shown in Figure 4. Similarly, the application is modeled as a graph, where each node t_i is annotated with the performance and energy consumption of the task when executed on each compatible library component and its deadline. The edges in the graph describe the data flow between the tasks and specify the required data bandwidth. The proposed approach employs a branch-and-bound algorithm to find a mapping of tasks to components. For each feasible mapping, the algorithm places the IP cores used in the design, and evaluates the metrics used in Equation 1. Finally, it selects the solution that minimizes the energy consumption while satisfying all the constraints.

3.2. Interconnection Network Design

Systems-on-polymer based on the FHE technology need to offer integrated sensing, local processing and energy efficient communication to support wearable applications and sensor-rich personalized computing experience. Hence, they need to integrate many rigid and flexible IP cores similar to the SoCs. SoC design experience shows that communication between these resources is critical for integration, system performance, power consumption, and validation [39, 46, 47]. Therefore, SoPs need energy-efficient interconnection networks that facilitate integrating many IP cores without introducing performance become bottlenecks.

Interconnection network design poses two major challenges. First, there is a need for reliable circuits and manufacturing solutions to interface rigid components with flexible substrates. This problem is addressed by novel circuits [31,48] and recently funded FHE Manufacturing Institute [2]. The second aspect, which is the focus of this section, is designing a system level communication network similar to NoCs [39]. Compared to SoCs, the FHE communication bandwidth is severely limited by three factors:

1. *Low integration density:* Modern CMOS technology offers 11 metal layers, less than 100 nm metal pitch, and switching speeds in the GHz range. Flexible and stretchable substrates, on the contrary, have limited number of layers, wire pitch is in the order of 10μm [20] and slower speed [7].
2. *Physical limitations:* Larger interconnect area decreases flexibility, since the elasticity of the interconnects is significantly smaller than that of the substrates. For example, Copper wires have a higher Young's modulus of elasticity of 128 GPa compared to Polyimide's elasticity of 7.5 GPa [49]. Ser-

pentine interconnects can be stretched, but they increase the effective wire width, hence the interconnect area.

3. *Limited circuit support:* Fast on-chip communication over long links ($\sim$ millimeters) is enabled by optimally inserted repeaters and advanced signaling techniques designed using CMOS technology [50]. These techniques may not be a viable option for flexible interconnects, which are significantly slower.

Consequently, the communication bandwidth on flexible substrates is significantly lower than on the chip. Furthermore, global interconnects on flexible substrates suffer from lower reliability, since they are prone to physical damage. Finally, the composition of flexible and rigid resources is expected to evolve as the flexible electronics technology advances. Hence, the interconnection networks should adapt to the growing needs.

The potential of wireless inter- and intra-chip communication has already been demonstrated [51–54]. For example, on-chip wireless communication is shown to be more energy efficient than wired interconnects for distances over 7mm [52]. Wired communication over flexible and stretchable substrate are even more constrained than on die. Hence, short-range wireless communication is a promising alternative. For example, wireless communication becomes more energy efficient as the data rate increases, since the cost of the voltage-controlled oscillator and receiver power are amortized over larger data rates [55]. Therefore, wired communication may be preferred at low data rates, while wireless communication becomes more energy-efficient as the data rate increases [56]. The observation can lead to interesting design and run-time optimization approaches, since data rate requirements are typically diverse. For instance, the communication bandwidth requirements of sensors are typically low ($< MB/s$), while the memory and display interfaces are very demanding. Furthermore, communication throughput shows significant variations over time as a function of the workload. Consequently, a hybrid network of wired and wireless links, as well as dynamically switching between the two may prove to be optimal. This optimization requires *accurate analysis* of different signaling techniques over stretchable and flexible interconnects. The results of this analysis can be used to decide whether a pure wired, pure wireless, or a hybrid combination is best suited for the application at hand.

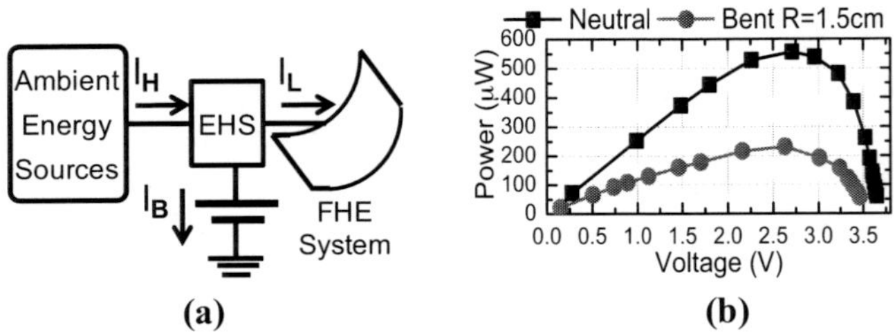

Figure 5. (a) A sample energy harvesting system (EHS) that can power FHE device and charge a flexible battery based on operating conditions. (b) Power generated by a 12.7 mm × 64 mm PV-cell when it is flat (neutral) and bent with a radius of curvature $R = 1.5$ cm.

3.3. Energy Harvesting and Runtime Management

Energy harvesting and dynamic power management are crucial for the practicality of FHE due to weight and flexibility constraints which prohibit large batteries. Major components of an energy harvesting system are the energy source, storage, harvesting circuit and harvesting aware energy management [57], as shown in Figure 5.

Energy sources: Photovoltaic (PV) cells provide 100 μW/cm^2 (indoor) – 100 mW/cm^2 (outdoor) power [57,58]. Moreover, they are one of the most mature areas of flexible electronics [15,59,60], and suit well to wearable applications. Similarly, flexible thermoelectric generators (TEG) provide 3.8 mW/cm^2 at $\Delta T = 50$ K [61] and piezoelectric generator generates up to 101 μW/cm^3 with 1 kgf [62].

When the PV-cell is flat, the angle between the cell and incident light is uniform across the surface. However, the incident angle at each infinitesimal point of a flexible PV-cell changes when the PV-cell is bent. As a result, energy harvested by flexible PV-cells can degrade significantly with bending, as we demonstrate empirically in Figure 5. We can derive the harvested energy under bending normalized to the flat cell as a function of the elevation angle of the sun α, the inclination angle of the device β, the length of PV-cell L, and the radius of curvature R [33]. For instance, the energy harvested at a bending radius of R

normalized to a flat PV-cell of length L is given as:

$$\text{Normalized Energy } (\mathrm{N_E}) = \frac{R[cos(\alpha+\beta-\frac{L}{2R})-cos(\alpha+\beta+\frac{L}{2R})]}{Lsin(\alpha+\beta)},$$
$$\lim_{R\to\infty} N_E = 1 \text{ (flat)}. \qquad (3)$$

We note that as the radius of curvature approaches infinity, the harvested energy converges to that of a flat cell. This model can be used in conjunction with maximum power point tracking algorithms to optimize the harvested energy.

Energy Storage and Harvesting Circuits: The amount of harvested energy varies as a function of the operating voltage, as shown in Figure 5(b). Furthermore, partial shading affect can introduce local maximum points on the voltage-power curve [33]. Therefore, operating at the maximum power point is important to prolong the service time of FHE systems running on harvested energy. Maximum power point tracking (MPPT) algorithms maximizes the harvested energy from the energy sources [63–65]. As bending changes the voltage-current characteristics, MPPT algorithms need to take dynamic shape changes into account. For instance, a MPPT algorithm based on the flexibility harvests up to 25% more energy compared to a MPPT algorithm without a flexibility consideration [33].

Power generated by ambient energy sources vary as a function of the environmental conditions, such as time of the day and temperature. Similarly, the power consumption of a FHE system fluctuating with workload variations. Consequently, the intermittent nature of the energy sources and the mismatch between the production and consumption necessitate an energy storage element, such as a battery and super capacitance [57]. Rechargeable flexible batteries [16] are good energy storage candidates for a FHE system in order not to sacrifice the flexibility. At the same time, it is important to minimize the storage requirements and completely eliminate it eventually with the help of optimal energy allocation [66], non-volatile logic [67] and novel application areas [68].

4. FHE Prototype and Potential Application Areas

Despite the phenomenal advances in the computational power and functionality of electronic systems, our interactions with them have largely been limited to simple control panels, keyboard, mouse and display. Moreover, devices based on current technology are not amenable to wearable applications, since they require frequent charging and are rigid unlike human skin and clothing. Consequently, these systems either rely critically on close human guidance or operate almost independently from the user. An exemplar technology that integrated tightly into our lives is the smartphone. However, the term "smartphone" is a misnomer, since it has fundamentally no intelligence to understand its user. The users still have to type, touch or speak (to some extent) to express their intentions in a form accessible to the phone. Hence, intelligent decision making is still almost entirely a human task.

In contrast to traditional technology, flexible hybrid electronics technology facilitates bendable and stretchable wearable systems at a low cost. Furthermore, wearable sensors can enable processing physiological signals and understand the user intent [69]. In addition, need for batteries can dramatically reduce or even vanish with the help the energy harvested from flexible PV-cells, motion and body heat. Therefore, FHE can support the IoT vision of a massive network of addressable smart objects by enabling arbitrarily shaped devices with integrated sensing, processing and communication capabilities [70–72]. Similarly, SoPs with physiological, biochemical, and motion sensing capabilities can be used as the nodes of wireless body area networks (WBAN), which interconnect a variety of sensing/actuating nodes in or on the body through an energy-aware wireless network [73–76].

One of the earliest and high-impact application areas of FHE is in health and activity monitoring. Wearable sensors have already started to find revolutionary applications to monitor human activity anywhere and anytime rather than being limited to a clinical environment [77, 78]. Furthermore, inertial measurement unit (IMU), magnetometer sensors and integrated radio enable low-cost indoor localization and tracking using handheld and wearable devices [79]. For example, wearable activity monitors, such as the Fitbit activity tracker [80], Google Fit [81] and numerous mobile apps have become widespread.

FHE technology can also be used for gesture recognition and human-machine communication. Wearable sensors embedded into electronic textiles [82, 83] and biosensors [69] enable accurate and real-time tracking of human motion as well as electrical biosignals, such as Electroencephalogram (EEG), Electromyogram (EMG) and Electrooculography (EOG). For instance,

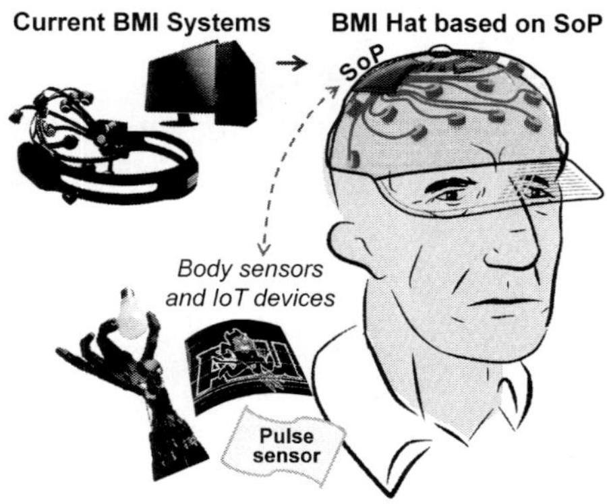

Figure 6. A brain-machine interface (BMI) implemented using a system-on-polymer integrated on a hat. A combination of EEG and motion sensing, in situ processing and wireless communication can enable agents that can decode human intent and lead to symbiotic human-machine relationship.

an early example of wearable gesture control is provided by the commercial Myo armband [84]. Similarly, recently introduced wearable finger sensors provide natural way of human-machine interaction through gesture recognition [85]. A combination of health monitoring, gesture recognition and human-machine communication can enable assistive technologies for the physically challenged and elderly population. According to the Annual World Report on Disability, 15% of the world's population lives with a disability, while 110 to 190 million of these people have significant difficulties in functioning [86]. Furthermore, the number of people over 65 years old is expected to double by 2050 [24, 25]. Significant amount of this population can substantially benefit from assistive technologies [87, 88]. For example, around 5.6 million, out of 7.6 million people with hearing impairment, could use a hearing aid according to U.S. Census Bureau [24]. We can significantly improve the quality of life of this population by providing accessibility to smart cyber-physical systems (CPS) that process sensory inputs and assist with everyday tasks [26, 89].

In the longer term, the FHE technology can bring a life-changing experience by transforming electronic devices from passive tools to self-sustained agents that can understand the human physiology, know what their user wants [90], as illustrated in Figure 6. As a result, it can deliver a richer experience than current smartphones by enriching them with more sensing and energy harvesting capabilities. Hence, it can drive the next big leap forward in the form factor design, similar to the shift from desktop and laptop computers to hand-held devices, as illustrated in Figure 1. For example, we envision large screen smartphones that can be rolled into a pen or folded.

Concrete prototypes enable us to validate the design methodologies and illustrate potential applications besides driving progress towards the longer-term goal (i.e., devices that can *understand and assist* their users), To this end, we implemented prototype shown in Figure 7 as a proof-of-concept [32]. It integrates a gyroscope, accelerometer, EMG/ECG processing circuitry, a TI-CC2650 microcontroller (MCU) [91], and printed antenna on a flexible Polyimide substrate. The MCU supports both BLE and Zigbee wireless communication protocols. The maximum supported data rate is 192 kbps, and the average power consumption at 1.44 kbps throughput is 12.2 mW. It occupies 1.5 in$\times$1.5 in and weighs 50 mg including the programming interfaces. Small area and low weight allow us to attach a dozen devices directly to the skin or clothing. We use this prototype for human motion processing, gesture recognition and assistive navigation [92, 93].

Figure 7. Our FHE prototype.

Conclusion and Future Prospects

Emerging flexible hybrid electronics technology integrates rigid silicon chips and printed electronics to bridge the gap between today's complex systems and flexible electronics. Physical flexibility and stretchability expands the design space into an uncharted dimension, which introduces intricate trade-offs with the traditional power, performance and area metrics. Therefore, there is a need for flexibility-aware design methodologies and tools tailored for flexible hybrid electronics. By complementing the advances in the manufacturing technology, these methodologies can be an enabler for ubiquitous devices for wearable computing, internet-of-things, body sensors, and medical applications. In the longer term, the FHE technology can drive the next big leap forward in the form factor design, similar to the shift from desktop and laptop computers to hand-held devices.

References

[1] W. S. Wong and A. Salleo, *Flexible Electronics: Materials and Applications*. Springer, 2009, vol. 24.

[2] Flexible Hybrid Electronics Manufacturing Institute. [Online] http://www.nextflex.us/, accessed 11 August 2017.

[3] T. Sekitani, U. Zschieschang, H. Klauk, and T. Someya, "Flexible organic transistors and circuits with extreme bending stability," *Nature Materials*, vol. 9, no. 12, pp. 1015–1022, 2010.

[4] A. Nathan, A. Ahnood, M. T. Cole, S. Lee, Y. Suzuki, P. Hiralal, F. Bonaccorso, T. Hasan, L. Garcia-Gancedo, A. Dyadyusha, S. Haque, P. Andrew, S. Hofmann, J. Moultrie, D. Chu, A. J. Flewitt, A. C. Ferrari, M. J. Kelly, J. Robertson, G. A. J. Amaratunga, and W. I. Milne, "Flexible Electronics: The Next Ubiquitous Platform," *Proc. of IEEE*, vol. 100, no. Special Centennial Issue, pp. 1486–1517, 2012.

[5] S. J. Shivakumar, Ed., *Flexible Electronics for Security, Manufacturing, and Growth in the United States: Summary of a Symp.* Washington DC: The National Academies Press, 2013.

[6] T.-C. J. Huang, J.-L. Huang, and C. Kwang-Ting, “Design, Automation, and Test for Low-Power and Reliable Flexible Electronics,” *Found. and Trends in EDA*, vol. 9, no. 2, pp. 99–210, 2015.

[7] R. Street, T. Ng, D. Schwartz, G. Whiting, J. Lu, R. Bringans, and J. Veres, “From Printed Transistors to Printed Smart Systems,” *Proceedings of the IEEE*, vol. 103, no. 4, pp. 607–618, April 2015.

[8] J. Noh, M. Jung, Y. Jung, C. Yeom, M. Pyo, and G. Cho, “Key Issues With Printed Flexible Thin Film Transistors and Their Application in Disposable RF Sensors,” *Proceedings of the IEEE*, vol. 103, no. 4, pp. 554–566, April 2015.

[9] T.-C. Huang, J.-L. Huang, and K.-T. Cheng, “Robust Circuit Design for Flexible Electronics,” *IEEE Design and Test of Computers*, vol. 28, no. 6, pp. 8–15, 2011.

[10] H. E. Katz and J. Huang, “Thin-Film Organic Electronic Devices,” *Annual Review of Materials Research*, vol. 39, pp. 71–92, 2009.

[11] ASU. Flexible Display Center, [Online] https://flexdisplay.asu.edu/, accessed 17 July 2016.

[12] T. Someya, Y. Kato, T. Sekitani, S. Iba, Y. Noguchi, Y. Murase, H. Kawaguchi, and T. Sakurai, “Conformable, Flexible, Large-area Networks of Pressure and Thermal Sensors with Organic Transistor Active Matrixes,” in *Proc. of the National Academy of Sciences of the United States of America*, vol. 102, no. 35, 2005, pp. 12 321–12 325.

[13] F. Zhang, M. Johansson, M. R. Andersson, J. C. Hummelen, and O. Inganäs, “Polymer Photovoltaic Cells with Conducting Polymer Anodes,” *Advanced Materials*, vol. 14, no. 9, pp. 662–665, 2002.

[14] H. Kang, H. Park, Y. Park, M. Jung, B. C. Kim, G. Wallace, and G. Cho, “Fully Roll-to-Roll Gravure Printable Wireless (13.56 [emsp14] MHz) Sensor-Signage Tags for Smart Packaging,” *Scientific Reports*, vol. 4, no. 5387, 2014.

[15] M. Koo, K.-I. Park, S. H. Lee, M. Suh, D. Y. Jeon, J. W. Choi, K. Kang, and K. J. Lee, "Bendable Inorganic Thin-Film Battery for Fully Flexible Electronic Systems," *Nano Letters*, vol. 12, no. 9, pp. 4810–4816, 2012.

[16] J. MacKenzie and C. Ho, "Perspectives on Energy Storage for Flexible Electronic Systems," *Proceedings of the IEEE*, vol. 103, no. 4, pp. 535–553, April 2015.

[17] A. Sou, S. Jung, E. Gili, V. Pecunia, J. Joimel, G. Fichet, and H. Sirringhaus, "Programmable Logic Circuits for Functional Integrated Smart Plastic Systems," *Elsevier, Organic Electronics*, vol. 15, no. 11, pp. 3111–3119, 2014.

[18] American Semiconductor Inc. FleX™ Silicon-on-Polymer™. http://www.americansemi.com/FleX.html, accessed 20 June 2016.

[19] R. L. Chaney, D. G. Wilson, D. R. Hackler, K. J. DeGregorio, and D. E. Leber, "New Silicon Frontiers: Physically Flexible System-On-A-Chip," in *Intl. Workshop on Microelectronics and Electron Devices*, 2017, pp. 1–4.

[20] M. Gonzalez, F. Axisa, M. V. Bulcke, D. Brosteaux, B. Vandevelde, and J. Vanfleteren, "Design of Metal Interconnects for Stretchable Electronic Circuits," *Elsevier Microelectronics Reliability*, vol. 48, no. 6, pp. 825–832, 2008.

[21] S. Xu *et al.*, "Soft Microfluidic Assemblies of Sensors, Circuits, and Radios for the Skin," *Science*, vol. 344, no. 6179, pp. 70–74, 2014.

[22] J. Sheats, "Truly wearable electronics," in *Presented at the FlexTech Alliance Workshop: Flexible Hybrid Electronics - Challenges and Solutions*, July 2014.

[23] R. L. Chaney, D. R. Hackler, D. G. Wilson, and B. N. Meek, " FleX™ Silicon-on-Polymer™: Flexible (Pliable) ICs from Commercial Foundry Processes," in *Proc. of GOMACTech (Government Microcircuit Applications and Critical Technology) Conf.*, March 2013.

[24] M. W. Brault. Americans With Disabilities: 2010. U.S. Census Bureau, Current Population Reports. Issued July 2012, https://www.census.gov/prod/2012pubs/p70-131.pdf, accessed 20 July 2015.

[25] L. A. Jacobsen, M. Kent, M. Lee, and M. Mather, "America's Aging Population," *Population Bulletin*, vol. 66, no. 1, 2011, population Reference Bureau, http://www.prb.org/pdf11/aging-in-america.pdf, accessed 20 July 2015.

[26] M. C. Domingo, "An Overview of the Internet of Things for People with Disabilities," *J. of Network and Computer Applications*, vol. 35, no. 2, pp. 584–596, 2012.

[27] B. Latré, B. Braem, I. Moerman, C. Blondia, and P. Demeester, "A Survey On Wireless Body Area Networks," *Wireless Networks*, vol. 17, no. 1, pp. 1–18, 2011.

[28] P. Summary, "WTEC Workshop on Int. R&D in Flexible Hybrid Electronics," in *World Technology Evaluation Center, Inc. (WTEC)*, June 2009, http://www.wtec.org/flex/documents/AllContent-FLEX-finalWS-Proceedings-30June2009.pdf, accessed 16 July 2015.

[29] S. Gowrisanker, M. Quevedo-Lopez, H. Alshareef, B. Gnade, S. Venugopal, R. Krishna, K. Kaftanoglu, and D. Allee, "A Novel Low Temperature Integration of CMOS Devices on Flexible Substrates," *Organic Electronics*, vol. 10, no. 7, pp. 1217–1222, 2009.

[30] Y. Hu, L. Huang, W. S. Rieutort-Louis, J. Sanz-Robinson, J. C. Sturm, S. Wagner, and N. Verma, "A Self-powered System for Large-scale Strain Sensing by Combining CMOS ICs with Large-area Electronics," *IEEE Journal of Solid-State Circuits*, vol. 49, no. 4, pp. 838–850, 2014.

[31] T. Moy, W. Rieutort-Louis, Y. Hu, L. Huang, J. Sanz-Robinson, J. C. Sturm, S. Wagner, and N. Verma, "Thin-film Circuits for Scalable Interfacing between Large-area Electronics and CMOS ICs," in *Device Research Conference*, 2014, pp. 271–272.

[32] U. Gupta, J. Park, H. Joshi, and U. Y. Ogras, "Flexibility-aware Systems on Polymer: Concept to Prototype," *IEEE Trans. on Multi Scale Computing Systems*, vol. 3, no. 1, pp. 36–49, 2017.

[33] J. Park, H. Joshi, H. G. Lee, S. Kiaei, and U. Y. Ogras, "Flexible PV-cell Modeling for Energy Harvesting in Wearable IoT Applications," *ACM Transactions on Embedded Computing Systems*, 2017.

[34] L.-T. Wang, Y.-W. Chang, and K.-T. T. Cheng, *Electronic Design Automation: Synthesis, Verification, and Test.* Morgan Kaufmann, 2009.

[35] R. Marculescu, U. Y. Ogras, and N. H. Zamora, "Computation and Communication Refinement for Multiprocessor SoC Design: A System-Level Perspective," *ACM Transactions on Design Automation of Electronic Systems (TODAES)*, vol. 11, no. 3, pp. 564–592, 2006.

[36] G. D. Micheli, *Synthesis and Optimization of Digital Circuits.* McGraw-Hill Higher Education, 1994.

[37] N. A. Sherwani, *Algorithms for VLSI physical design automation.* Springer Science & Business Media, 2012.

[38] A. Sangiovanni-Vincentelli and G. Martin, "Platform-Based Design and Software Design Methodology for Embedded Systems," *IEEE Design & Test of Computers*, vol. 18, no. 6, pp. 23–33, 2001.

[39] U. Y. Ogras and R. Marculescu, *Modeling, Analysis and Optimization of Network-on-Chip Communication Architectures.* Springer, 2013, vol. 184.

[40] D. R. Askeland and P. P. Phulé, *The Science and Engineering of Materials.* Springer, 2006.

[41] K. Feng and Z.-C. Shi, *Mathematical Theory of Elastic Structures.* Springer Science & Business Media, 2013.

[42] COMSOL Multiphysics©. The Platform for Physics-Based Modeling and Simulation. https://www.comsol.com/comsol-multiphysics, 7 July 2015.

[43] M. R. Garey and D. S. Johnson, *Computers and Intractability.* New York, NY: W. H. Freeman & Co., 2002, vol. 29.

[44] J. Hu and R. Marculescu, "Energy-And Performance-Aware Mapping for Regular Noc Architectures," *IEEE Trans. on Computer-Aided Design of Integrated Circuits and Systems*, vol. 24, no. 4, pp. 551–562, 2005.

[45] G. Bhat, U. Gupta, N. Tran, J. Park, S. Ozev, and U. Y. Ogras, "Multi-Objective Design Optimization for Flexible Hybrid Electronics," in *Proc. of Int. Conf. on Computer-Aided Design*, Nov. 2016, http://elab.engineering.asu.edu/wp-content/uploads/2016/07/BhatICCAD2016.pdf, 18 July 2016.

[46] L. Benini and G. De Micheli, "Networks on Chip: A New Paradigm for Systems on Chip Design," in *Proc. of Conf. on Design, Automation and Test in Europe*, 2002, pp. 418–419.

[47] A. Jantsch, H. Tenhunen *et al.*, *Networks-on-Chip.* Springer, 2003, vol. 396.

[48] N. Verma, Y. Hu, L. Huang, W. S. Rieutort-Louis, J. S. Robinson, T. Moy, B. Glisic, S. Wagner, and J. C. Sturm, "Enabling Scalable Hybrid Systems: Architectures for Exploiting Large-area Electronics in Applications," *Proceedings of the IEEE*, vol. 103, no. 4, pp. 690–712, 2015.

[49] J. Dolbow and M. Gosz, "Effect of Out-Of-Plane Properties of a Polyimide Film on the Stress Fields in Microelectronic Structures," *Elsevier Mechanics of Materials*, vol. 23, no. 4, pp. 311–321, 1996.

[50] H. Cho, P. Kapur, and K. C. Saraswat, "Power Comparison Between High-Speed Electrical and Optical Interconnects for Interchip Communication," *OSA Jrnl. of Lightwave Tech.*, vol. 22, no. 9, p. 2021, 2004.

[51] M.-C. F. Chang, I. Verbauwhede, C. Chien, Z. Xu, J. Kim, J. Ko, Q. Gu, and B.-C. Lai, "Advanced RF/Baseband Interconnect Schemes for Inter-and Intra-ULSI Communications," *IEEE Tran. on Electron Devices*, vol. 52, no. 7, pp. 1271–1285, 2005.

[52] S. Deb, A. Ganguly, P. P. Pande, B. Belzer, and D. Heo, "Wireless NoC as Interconnection Backbone for Multicore Chips: Promises and Challenges," *IEEE Jrnl. on Emerging and Selected Topics in Circuits and Systems*, vol. 2, no. 2, pp. 228–239, 2012.

[53] D. Zhao and Y. Wang, "MTNET: Design and optimization of a wireless SOC test framework," in *Proc. of Intl. SOC Conf.*, 2006, pp. 239–242.

[54] A. Ganguly, K. Chang, S. Deb, P. P. Pande, B. Belzer, and C. Teuscher, "Scalable Hybrid Wireless Network-on-Chip Architectures for Multicore Systems," *IEEE Trans. on Computers*, vol. 60, no. 10, pp. 1485–1502, 2011.

[55] C. W. Byeon, C. H. Yoon, and C. S. Park, "A 67-mW 10.7-Gb/s 60-GHz OOK CMOS Transceiver for Short-Range Wireless Communications," *IEEE Trans. on Microwave Theory and Techniques*, vol. 61, no. 9, pp. 3391–3401, 2013.

[56] U. Gupta and U. Y. Ogras, "Extending Networks from Chips to Flexible and Stretchable Electronics," in *Proc. of Networks-on-Chip Symp.*, Sept. 2016, http://elab.engineering.asu.edu/wp-content/uploads/2016/07/GuptaNOCS2016.pdf, 18 July 2016.

[57] V. Raghunathan, A. Kansal, J. Hsu, J. Friedman, and M. Srivastava, "Design Considerations for Solar Energy Harvesting Wireless Embedded Systems," in *Proc. of Intl. Symp. on Information Processing in Sensor Networks*, 2005, p. 64.

[58] A. Valenzuela. (2008) Energy Harvesting for No-Power Embedded Systems. http://focus.ti.com/graphics/mcu/ulp/energy_harvesting_embedded_systems_using_msp430.pdf, accessed 19 July 2016.

[59] S. Venugopal, D. Allee, M. Quevedo-Lopez, B. Gnade, E. Forsythe, and D. Morton, "Flexible Electronics: What Can It Do? What Should It Do?" in *Proc. of Reliability Physics Symp.*, vol. 2010, 2010, pp. 644–649.

[60] M. Pagliaro, R. Ciriminna, and G. Palmisano, "Flexible Solar Cells," *ChemSusChem*, vol. 1, no. 11, pp. 880–891, 2008.

[61] S. J. Kim, J. H. We, and B. J. Cho, "A Wearable Thermoelectric Generator Fabricated on a Glass Fabric," *Energy & Environmental Science*, vol. 7, no. 6, pp. 1959–1965, 2014.

[62] P. G. Kang, B. K. Yun, K. D. Sung, T. K. Lee, M. Lee, N. Lee, S. H. Oh, W. Jo, H. J. Seog, C. W. Ahn, I. W. Kim, and J. H. Jung, "Piezoelectric Power Generation of Vertically Aligned Lead-free (K,Na)NbO_3 Nanorod Arrays," *RSC Advances*, vol. 4, no. 56, pp. 29 799–29 805, 2014.

[63] D. Brunelli, L. Benini, C. Moser, and L. Thiele, "An Efficient Solar Energy Harvester for Wireless Sensor Nodes," in *Proceedings of the Design, Automation and Test in Europe*, 2008, pp. 104–109.

[64] M. A. Elgendy, B. Zahawi, and D. J. Atkinson, "Assessment of Perturb and Observe MPPT Algorithm Implementation Techniques for PV Pumping Applications," *IEEE Transactions on Sustainable Energy*, vol. 3, no. 1, pp. 21–33, 2012.

[65] A. Safari and S. Mekhilef, "Simulation and Hardware Implementation of Incremental Conductance MPPT with Direct Control Method using Cuk Converter," *IEEE Transactions on Industrial Electronics*, vol. 58, no. 4, pp. 1154–1161, 2011.

[66] G. Bhat, J. Park, and U. Y. Ogras, "Near-Optimal Energy Allocation for Self-Powered Wearable Systems," in *Proceedings of International Conference on Computer Aided Design*, 2017.

[67] S. C. Bartling, S. Khanna, M. P. Clinton, S. R. Summerfelt, J. A. Rodriguez, and H. P. McAdams, "An 8MHz 75μA/MHz Zero-Leakage Non-Volatile Logic-based Cortex-M0 MCU SoC Exhibiting 100% Digital State Retention at VDD= 0V with< 400ns Wakeup and Sleep Transitions," in *IEEE Intl. Solid-State Circuits Conference Digest of Technical Papers (ISSCC)*, 2013, pp. 432–433.

[68] Y. Wang, Y. Liu, C. Wang, Z. Li, X. Sheng, H. G. Lee, N. Chang, and H. Yang, "Storage-less and Converter-less Photovoltaic Energy Harvesting with Maximum Power Point Tracking for Internet of Things," *IEEE Trans. on Computer-Aided Design of Integrated Circuits and Systems*, vol. 35, no. 2, pp. 173–186, 2016.

[69] S.-T. Han, H. Peng, Q. Sun, S. Venkatesh, K.-S. Chung, S. C. Lau, Y. Zhou, and V. A. L. Roy, "An Overview of the Development of Flexible Sensors," *Advanced Materials*, p. 1700375, July 2017.

[70] D. Guinard, V. Trifa, S. Karnouskos, P. Spiess, and D. Savio, "Interacting with the Soa-Based Internet Of Things: Discovery, Query, Selection, and On-Demand Provisioning Of Web Services," *IEEE Trans. on Services Computing*, vol. 3, no. 3, pp. 223–235, 2010.

[71] A. Bassi, M. Bauer, M. Fiedler, T. Kramp, R. Kranenburg, S. Lange, and S. Meissner, *Enabling Things To Talk: Designing Iot Solutions With The Iot Architectural Reference Model*, 1st ed. Springer, 2013.

[72] J. Gubbi, R. Buyya, S. Marusic, and M. Palaniswami, "Internet of Things (Iot): A Vision, Architectural Elements, and Future Directions," *Future Generation Computer Systems*, vol. 29, no. 7, pp. 1645–1660, 2013.

[73] P. Bonato, "Wearable Sensors/Systems And Their Impact On Biomedical Engineering," *IEEE Engineering in Medicine and Biology Magazine*, vol. 22, no. 3, pp. 18–20, 2003.

[74] F. Zhang, Y. Zhang, J. Silver, Y. Shakhsheer, M. Nagaraju, A. Klinefelter, J. Pandey, J. Boley, E. Carlson, A. Shrivastava, B. Otis, and B. H. Calhoun, "A batteryless 19μW MICS/ISM-band energy harvesting body area sensor node SoC," in *Proc. of Intl. Solid-State Circuits Conference Digest of Technical Papers*, 2012, pp. 298–300.

[75] H. Cao, V. Leung, C. Chow, and H. Chan, "Enabling Technologies For Wireless Body Area Networks: A Survey And Outlook," *IEEE Communications Magazine*, vol. 47, no. 12, pp. 84–93, 2009.

[76] M. Chen, S. Gonzalez, A. Vasilakos, H. Cao, and V. C. Leung, "Body Area Networks: A Survey," *Mobile Networks and Applications*, vol. 16, no. 2, pp. 171–193, 2011.

[77] S. C. Mukhopadhyay, "Wearable Sensors for Human Activity Monitoring: A Review," *IEEE Sensors Journal*, vol. 15, no. 3, pp. 1321–1330, 2015.

[78] A. Nag, S. C. Mukhopadhyay, and J. Kosel, "Wearable Flexible Sensors: A Review," *IEEE Sensors Journal*, vol. 17, no. 13, pp. 3949–3960, July 2017.

[79] M. G. Jadidi, M. Patel, J. V. Miro, G. Dissanayake, J. Biehl, and A. Girgensohn, "Calibrationless Radio-Inertial Localization and Tracking Systems using Smart Handheld Devices," *arXiv preprint arXiv:1706.05569*, 2017.

[80] Fitbit Activity Tracker. [Online] https://www.fitbit.com/, accessed 13 August 2017.

[81] Google Fit. [Online] https://www.google.com/fit/, accessed 20 August 2017.

[82] M. Stoppa and A. Chiolerio, "Wearable electronics and smart textiles: a critical review," *Sensors*, vol. 14, no. 7, pp. 11 957–11 992, 2014.

[83] D. Marculescu *et al.*, "Electronic textiles: A platform for pervasive computing," *Proceedings of the IEEE*, vol. 91, no. 12, pp. 1995–2018, 2003.

[84] Myo Gesture Control Armband, Wearable Technology by Thalmic Labs. [Online] https://www.myo.com/, accessed 13 August 2017.

[85] Project Jacquard. [Online] https://atap.google.com/jacquard/, accessed 20 August 2017.

[86] World Report on Disability. [Online] http://www.who.int/disabilities/world_report/2011/en/, accessed 20 August 2017.

[87] E. C. Bouck, *Advances in Special Education.* Emerald Group Publishing Limited, 2010, vol. 20, ch. 6. Technology and Students with Disabilities: Does it Solve All the Problems?, pp. 91–104.

[88] K. E. Forgrave, "Assistive Technology: Empowering Students with Learning Disabilities," *The Clearing House: A Journal of Educational Strategies, Issues and Ideas*, vol. 75, no. 3, pp. 122–126, 2002, http://www.tandfonline.com/doi/pdf/10.1080/00098650209599250, accessed 20 July 2015.

[89] S. Shahrestani, *Internet of Things and Smart Environments: Assistive Technologies for Disability, Dementia, and Aging*. Springer, 2017.

[90] B. Mahmoudi and J. C. Sanchez, "A symbiotic brain-machine interface through value-based decision making," *PloS one*, vol. 6, no. 3, p. e14760, 2011.

[91] Texas Instruments Inc. CC-2650 Microcontroller. http://www.ti.com/product/CC2650, accessed 30 June 2016.

[92] C. S. Geyik, A. Dutta, U. Y. Ogras, and D. W. Bliss, "Decoding Human Intent Using a Wearable System and Multi-Modal Sensor Data," in *IEEE Asilomar Conf. on Signals, Systems and Computers*, 2016, pp. 846–850.

[93] A. Porter, M. Muztoba, and U. Y. Ogras, "Human-machine Communication for Assistive IoT Technologies," in *Proc. of the Intl. Conf. on Hardware/Software Codesign and System Synthesis*, 2016, p. 31.

Biographical Sketch

Umit Y. Ogras, PhD

Affiliation: School of Electrical, Computer and Energy Engineering, Arizona State University, AZ, US

Education: PhD in Electrical and Comp. Engineering, December 2007, Carnegie Mellon University, Pittsburgh, PA, US

Business Address: 781 E Terrace Rd. ISTB4, Tempe AZ 85287-5804

Research and Professional Experience: Umit Y. Ogras received his PhD degree in Electrical and Computer Engineering from Carnegie Mellon University, Pittsburgh, PA, in 2007. From 2008 to 2013, he worked as a Research Scientist at the Strategic CAD Laboratories, Intel Corporation. He is currently an Assistant Professor at the School of Electrical, Computer and Energy

Engineering. Recognitions Dr. Ogras has received include Strategic CAD Labs Research Award, 2012 IEEE Donald O. Pederson Transactions on CAD Best Paper Award, 2011 IEEE VLSI Transactions Best Paper Award and 2008 EDAA Outstanding PhD Dissertation Award. His research interests include digital system design, embedded systems, flexible hybrid electronics, multicore architecture and electronic design automation with particular emphasis on multiprocessor systems-on-chip (MPSoC).

Professional Appointments:
08/2013 - Present: Assistant Professor, Arizona State University
06/2008 - 08/2013: Research Scientist, Intel Corporation
12/2007 - 06/2008: Post-Doctoral Research Associate, Carnegie Mellon University
06/2003 - 12/2007: Graduate Research Assistant, Carnegie Mellon University

Honors:
– NSF CAREER Award (2017);
– 2013 Intel Design and Technology Solution Division Recognition Award for enabling convergence on the on-chip interconnect of Xeon Server and Xeon Phi processors;
– 2012 Intel Strategic CAD Labs Research Award for developing technology for exploration of communication fabrics and driving adaptation by architecture teams;
– IEEE Council on Electronic Design Automation 2012 Donald O. Pederson Trans. On CAD Best Paper Award;
– IEEE Circuits and Systems Society 2011 VLSI Transactions Best Paper Award;
– 2008 Outstanding PhD Dissertation Award from the European Design Automation Association.

Publications from the Last 3 Years: The list of publications can be found at: https://elab.engineering.asu.edu/publications/

INDEX

A

B

C

D

E

F

G

H

I

M

N

O

P

R

S

T

U

V

W

M